Advanced
Core 2 for

Welcome to Advanced Mat...
your examination performa...
OCR Core 2 examination. I...
studied. Each chapter has th...
sub-heading gives the OCR...

The book contains scores of worked examples, each with clearly set-out steps to help solve the problem. You can then apply the steps to solve the Skills Check questions in the book and past exam questions at the end of each chapter. If you feel you need extra practice on any topic, you can try the Skills Check Extra exercises on the accompanying CD-ROM. At the back of this book there is a sample exam-style paper to help you test yourself befo... th...book.

Some of the questions in the book have a ⊙ symbol next to them. Th...
PowerPoint® solution (on the CD-ROM) that guides you through sugg...
the problem and setting out your answer clearly.

Using the CD-ROM

To use the accompanying CD-ROM simply put the disc in your CD-ROM drive, and the menu should appear automatically. If it doesn't automatically run on your PC:

1. Select the My Computer icon on your desktop.
2. Select the CD-ROM drive icon.
3. Select Open.
4. Select core2_for _ocr.exe.

016 916

If you don't have PowerPoint® on your computer you can download PowerPoint 2003 Viewer®. This will allow you to view and print the presentations. Download the viewer from http://www.microsoft.com

Pearson Education Limited
Edinburgh Gate
Harlow
Essex
CM20 2JE
England
www.longman.co.uk

First published 2005
ISBN   0 582 836522

Design by Ken Vail Graphic Design

Cover design by Raven Design

Typeset by Tech-Set, Gateshead

Printed in the U.K. by Scotprint, Haddington

The publisher's policy is to use paper manufactured from sustainable forests.

The publisher wishes to draw attention to the Single-User Licence Agreement at the back of the book.
Please read this agreement carefully before installing and using the CD-ROM.

We are grateful for permission from OCR to reproduce past exam questions. All such questions have a reference in the margin. OCR can accept no responsibility whatsoever for accuracy of any solutions or answers to these questions.

Every effort has been made to ensure that the structure and level of sample question papers matches the current specification requirements and that solutions are accurate. However, the publisher can accept no responsibility whatsoever for accuracy of any solutions or answers to these questions. Any such solutions or answers may not necessarily constitute all possible solutions.

1 Trigonometry

1.1 Sine and cosine rules

Use the sine and cosine rules in the solution of triangles.

The sine and cosine rules are used to find lengths and angles in a triangle.

To apply them, you need to label your triangle as follows:

Label the vertices with upper case letters, for example *A*, *B* and *C*. Then label the side opposite each vertex with the corresponding lower case letter, so side *a* is opposite angle *A*, side *b* is opposite angle *B* and side *c* is opposite angle *C*.

Note:
The triangle can be any size and shape.

Note:
A **vertex** is where two lines meet. The plural is **vertices**.

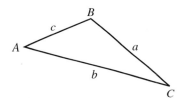

The sine rule

To find a **length**, use

$$\frac{a}{\sin A} = \frac{b}{\sin B} = \frac{c}{\sin C}$$

Format (1)

To find an **angle**, use

$$\frac{\sin A}{a} = \frac{\sin B}{b} = \frac{\sin C}{c}$$

Format (2)

Tip:
You can use the sine rule to find:
• a side when you know the angles in the triangle and another side
• an angle when you know two sides and the angle opposite one of them.

Tip:
To proceed, you must know one complete ratio.

Note:
You must learn the sine rule.

Example 1.1 In triangle *ABC*, *AC* is 3.2 cm, angle *ABC* is 35° and angle *BCA* is 82°. Find *AB*, giving your answer to the nearest mm.

Step 1: Draw a carefully labelled sketch and include all known measures.

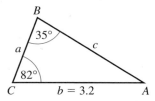

Step 2: Write down the sine rule in format (1) for finding a length.

Using the sine rule:

$$\frac{a}{\sin A} = \frac{b}{\sin B} = \frac{c}{\sin C}$$

Step 3: Substitute the known values.

$$\frac{a}{\sin A} = \frac{3.2}{\sin 35°} = \frac{c}{\sin 82°}$$

Step 4: Choose the two relevant ratios and solve the equation.

$$\frac{c}{\sin 82°} = \frac{3.2}{\sin 35°}$$

$$c = \frac{3.2 \times \sin 82°}{\sin 35°}$$

$$= 5.5247...$$

So *AB* = 5.5 cm (to the nearest mm)

Note:
You know length *b* and angle *B*. You are not asked anything about *a* and *A* so ignore the ratio involving them.

Note:
To give the answer to the nearest mm, you need to correct your value in cm to one decimal place.

1

Example 1.2 Find θ, giving your answer to the nearest degree.

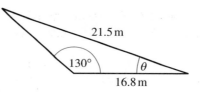

Step 1: Label the sketch carefully.

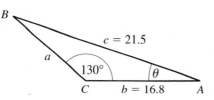

Note:
θ is angle A.

Step 2: Write down the sine rule in format (2) for finding an angle.

$$\frac{\sin A}{a} = \frac{\sin B}{b} = \frac{\sin C}{c}$$

Step 3: Substitute the known values.

$$\frac{\sin \theta}{a} = \frac{\sin B}{16.8} = \frac{\sin 130°}{21.5}$$

Step 4: Choose the two relevant ratios and solve the equation.

$$\frac{\sin B}{16.8} = \frac{\sin 130°}{21.5}$$

$$\sin B = \frac{16.8 \times \sin 130°}{21.5}$$

$$= 0.59858...$$

$$B = 36.76...°$$

$$\Rightarrow \quad \theta = 180° - (130° + 36.76...°)$$

$$= 13.23...°$$

$$= 13° \text{ (to the nearest degree)}$$

Note:
As you do not know a, you cannot use $\dfrac{\sin \theta}{a}$, but you *can* use the middle ratio to find angle B. You can then calculate θ.

Note:
$B = \sin^{-1}(0.59858...)$

Tip:
The sum of the angles in a triangle is 180°.

The cosine rule

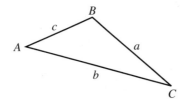

To find a **length**, for example a, use this formula for a^2:

$$a^2 = b^2 + c^2 - 2bc \cos A \qquad \textit{Format (1)}$$

Then square root to calculate a.

To find b, use

$$b^2 = a^2 + c^2 - 2ac \cos B$$

To find c, use

$$c^2 = a^2 + b^2 - 2ab \cos C$$

To find an **angle**, rearrange the formulae as follows:

$$\cos A = \frac{b^2 + c^2 - a^2}{2bc} \qquad \textit{Format (2)}$$

$$\cos B = \frac{a^2 + c^2 - b^2}{2ac}$$

$$\cos C = \frac{a^2 + b^2 - c^2}{2ab}$$

Tip:
You can use the cosine rule to find:
- the third side when you know two sides and the angle between them
- an angle when you know the three sides.

Note:
You will be given format (1) in the examination.
Make sure you can rearrange it to find an angle. Remember the patterns.

Example 1.3 In triangle ABC, $AC = 6.2$ cm, $BC = 8.3$ cm and angle $ACB = 42°$. Calculate AB, giving your answer correct to three significant figures.

Step 1: Draw a carefully labelled sketch and include all known measures.

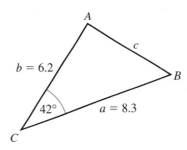

Tip:
The cosine rule is appropriate as you know two sides and the angle between them.

Step 2: Use the cosine rule to find the missing length.

By the cosine rule

$$c^2 = a^2 + b^2 - 2ab \cos C$$

$$= 8.3^2 + 6.2^2 - 2 \times 8.3 \times 6.2 \times \cos 42°$$

$$= 30.84...$$

$$c = \sqrt{30.84...} = 5.553...$$

$$AB = 5.55 \text{ cm (3 s.f.)}$$

Tip:
Find c^2 in one stage on your calculator; do not press $=$ until you have entered cos 42.

Example 1.4 In triangle XYZ, $XY = 6$ cm, $YZ = 8$ cm and $XZ = 12$ cm. Find the size of the largest angle in the triangle, giving your answer to the nearest $0.1°$.

Tip:
The largest angle is opposite the longest side.

Step 1: Draw a carefully labelled sketch and include all known measures.

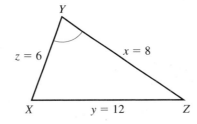

Step 2: Use the cosine rule to find the missing angle.

$$\cos Y = \frac{x^2 + z^2 - y^2}{2xz}$$

$$= \frac{8^2 + 6^2 - 12^2}{2 \times 8 \times 6}$$

$$= -0.458...$$

$$Y = 117.3° \text{ (nearest } 0.1°)$$

Calculator note:

There are several ways of entering this calculation. Make sure that the method you use is a correct one.

Here is an example:

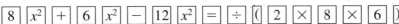

Tip:
Giving your answer to the nearest $0.1°$ is the same as approximating to one decimal place.

Tip:
The denominator must be enclosed in brackets here.

Use the area formula, $\Delta = \frac{1}{2}ab \sin C$.

When you know **two sides** and the **angle between them**, you can use this formula to calculate the **area** of a triangle.

$$\text{Area} = \tfrac{1}{2}ab \sin C$$

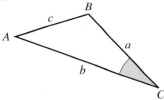

If you know angle A or angle B use one of the following formats:

$$\text{Area} = \tfrac{1}{2}bc \sin A$$
$$\text{Area} = \tfrac{1}{2}ac \sin B$$

Example 1.5 In triangle PQR, $PQ = 4.2$ cm, $QR = 6.3$ cm and angle $PQR = 130°$. Calculate the area of the triangle, giving your answer to two significant figures.

Step 1: Draw a carefully labelled sketch and include all known measures.

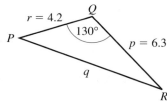

Step 2: Use the area formula.

$$\begin{aligned}
\text{Area} &= \tfrac{1}{2}\,pr \sin Q \\
&= \tfrac{1}{2} \times 6.3 \times 4.2 \times \sin 130° \\
&= 10.13\ldots \\
&= 10\ \text{cm}^2\ (2\ \text{s.f.})
\end{aligned}$$

Example 1.6 In triangle XYZ, $YZ = 5$ cm, $XY = 8$ cm and $XZ = 9$ cm.

Calculate the area of the triangle, giving your answer correct to two significant figures.

Step 1: Draw a carefully labelled sketch and include all known measures.

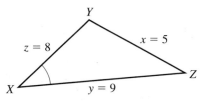

Step 2: Use the cosine rule to find an angle.

Calculate an angle, using the cosine rule.

$$\begin{aligned}
\cos X &= \frac{y^2 + z^2 - x^2}{2yz} \\
&= \frac{9^2 + 8^2 - 5^2}{2 \times 9 \times 8} \\
&= 0.833\ldots \\
X &= 33.557\ldots°
\end{aligned}$$

Step 3: Use the area formula.

$$\begin{aligned}
\text{Area} &= \tfrac{1}{2}\,yz \sin X \\
&= \tfrac{1}{2} \times 9 \times 8 \times \sin 33.557\ldots° \\
&= 19.89\ldots \\
&= 20\ \text{cm}^2\ (2\ \text{s.f.})
\end{aligned}$$

Tip:
It does not matter which angle you calculate.

Tip:
Do not approximate here but use the full display on the calculator in the next calculation.

Give answers to three significant figures unless requested otherwise.

1 Calculate *x*, correct to the nearest mm.

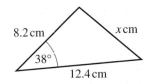

2 Calculate *y*.

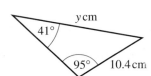

3 a Calculate angle *ABC*.

b Calculate the area of triangle *ABC*.

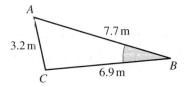

4 In triangle *QPR*, *QR* = 4 mm, *RP* = 5.5 mm,
angle *QPR* = 35°.

a Calculate angle *PQR*, given that it is acute.

b Calculate angle *QRP*.

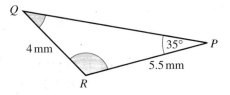

5 In triangle *ABC*, angle *BAC* = 15°, angle *ABC* = 140° and *AC* = 20.5 cm.

a Calculate length *BC*.   **b** Calculate the area of triangle *ABC*.

6 A triangle has two equal sides of length 6 cm and one of the angles in the triangle is 40°.

a Sketch two possible triangles.

b For each triangle, calculate
  i the area of the triangle,   **ii** the length of the third side.

 7 Calculate angle *QRS*.

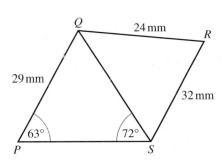

8 A circle, centre *O*, has radius 12 cm.
Angle *OPQ* = 37°. Calculate

a the length of the chord *PQ*,

b the area of triangle *OPQ*.

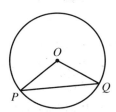

9 ABCD is a field in the shape of a quadrilateral.
$AB = 25$ m, $AD = 15$ m, $DC = 17$ m.
Angle $BAD = 45°$, angle $BCD = 62°$. Calculate

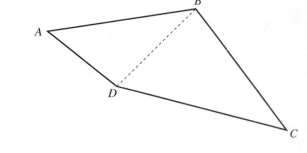

   a the length of the diagonal BD,

   b angle DBC,

   c angle BDC,

   d the area of the field.

 10 In triangle ABC, $AB = 8$ cm and $AC = 12$ cm.

   a The area of the triangle is 24 cm^2. Given that angle BAC is acute, calculate angle BAC and length BC.

   b Show that when angle BAC is $150°$, the area of triangle ABC is also 24 cm^2 and calculate the length BC in this case.

   c With the aid of diagrams, comment on your answers to parts **a** and **b**.

SKILLS CHECK **1A EXTRA** is on the CD

1.3 Degrees and radians

Understand the definition of a radian, and use the relationship between degrees and radians.

When the radius OP turns through an angle θ about O, the **sector** PQO is formed, with **arc length** PQ.

When the length of the arc is the same as the radius, the angle is 1 **radian** (1^c).

Note:
The symbol ^{c} stands for circular measure.

Converting between radians and degrees

Make sure that you learn the following:

π radians $= 180°$

To convert radians to degrees, multiply by $\dfrac{180}{\pi}$.

To convert degrees to radians, multiply by $\dfrac{\pi}{180}$.

Note:
In one complete turn of $360°$, the arc length is $2\pi r$ (2π lots of r), so the angle is 2π lots of 1 radian, that is 2π radians. This means that 2π radians $= 360°$.

Example 1.7 Write $40°$ in radians as a multiple of π.

Step 1: Multiply by $\dfrac{\pi}{180}$ and simplify, leaving π in your answer.

$40° = 40 \times \dfrac{\pi}{180}$ radians $= \dfrac{2}{9}\pi$ radians.

Tip:
When an angle is given as a multiple of π, the word radians or the symbol ^{c} is usually omitted.

You will find it helpful to learn these common conversions:

Radians	Degrees
$\frac{1}{6}\pi$	$30°$
$\frac{1}{4}\pi$	$45°$
$\frac{1}{3}\pi$	$60°$
$\frac{1}{2}\pi$	$90°$
π	$180°$
2π	$360°$
1^c	$57°$ (to nearest degree)
0.017^c (to 3 d.p.)	$1°$

Tip:
If you forget, use $180° = \pi$ radians and work them out yourself.

1.4 Arc length and sector area of a circle

Use the formulae $s = r\theta$ and $A = \frac{1}{2}r^2\theta$ for the arc length and sector area of a circle.

When θ is measured in radians:

arc length $\quad s = r\theta$

area of sector $\quad A = \frac{1}{2}r^2\theta$

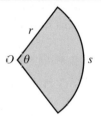

Tip:
Learn these and remember that θ must be in radians.

Example 1.8 The diagram shows a sector of a circle with centre O and radius r cm.

The angle POQ is 2 radians and the arc length PQ is 8 cm. Calculate

a the value of r,

b the area of sector POQ.

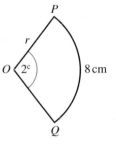

Step 1: Use the arc length formula to find the radius.

a $\quad\quad s = r\theta$

$\Rightarrow \quad 8 = r \times 2$

Step 2: Use the area formula.

$r = 4$

b $\quad A = \frac{1}{2}r^2\theta$

$= \frac{1}{2} \times 4^2 \times 2$

$= 16$

The area of the sector is 16 cm².

Example 1.9 A circle has centre O and radius 5 cm. Chord PQ has length 8 cm and angle POQ is θ radians.

Giving your answers correct to two significant figures, calculate

a the value of θ,

b the area, in cm², of the shaded segment.

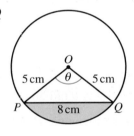

Step 1: Use the cosine rule to find θ.

a Using the cosine rule:

$$\cos\theta = \frac{5^2 + 5^2 - 8^2}{2\times5\times5}$$

$$= -0.28$$

$$\theta = 1.854\ldots$$

$$= 1.9 \text{ (2 s.f.)}$$

Tip:
Set your calculator to radian mode.

Tip:
Remember to give your answers to the requested accuracy.

Step 2: Find the area of sector POQ.

b $\quad$ Area of sector $POQ = \frac{1}{2}r^2\theta$

$$= \frac{1}{2} \times 5^2 \times 1.854\ldots$$

$$= 23.182\ldots \text{ cm}^2$$

Step 3: Find the area of triangle POQ.

Area of triangle $POQ = \frac{1}{2}r^2\sin\theta$

$$= \frac{1}{2} \times 5^2 \times \sin 1.854\ldots$$

$$= 12 \text{ cm}^2$$

Recall:
The formula for the area of a triangle (Section 1.2).

Step 4: Subtract the areas.

Area of the shaded segment

$$= \text{area of sector } POQ - \text{area of triangle } POQ$$

$$= 23.182\ldots - 12$$

$$= 11.182\ldots$$

$$= 11 \text{ cm}^2 \text{ (2 s.f.)}$$

Tip:
To maintain accuracy, do not use approximated answers found early in the solution but use the full display. It is useful to store values in the calculator memory.

Give answers to three significant figures unless requested otherwise.

1 Convert   **a** 280° to radians   **b** 1.5 radians to degrees.

2 Convert the following angles in radians to degrees.

    a $\frac{2}{3}\pi$ **b** $\frac{3}{4}\pi$ **c** $\frac{3}{2}\pi$ **d** $\frac{7}{12}\pi$

3 Convert these angles to radians, giving each angle in terms of π.

    a 45°   **b** 150°   **c** 330°   **d** 240°

4 *POQ* is a sector of a circle, centre *O*, radius 5 cm.
Angle *POQ* is 0.6 radians. Calculate

    a the length of arc *PQ*,   **b** the perimeter of the sector,

    c the area of triangle *POQ*,   **d** the area of sector *POQ*.

5 *AOB* is a sector of a circle, centre *O* and radius 10.4 cm. The arc length *AB* is 12.48 cm.

    a Find angle *AOB*, in radians.   **b** Find the area of sector *AOB*.

 6 A sector of a circle has area 27 cm² and radius 6 cm.

    a Calculate the angle of the sector, in radians.

    b Calculate the perimeter of the sector.

7 A circle, with centre *O*, has radius 8 cm.
A chord intersects the circle at *P* and *Q* and
angle *POQ* is θ radians, where θ is acute.

The area of triangle *POQ* is 24 cm². Find

    a the value of θ,

    b the area of sector *POQ*,

    c the area of the shaded segment.

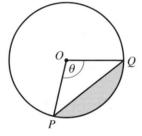

 8 *PQ* is an arc of a circle, centre *A* and radius 10 cm. *BC* is an arc of a circle, centre *A* and radius 7 cm.
The size of angle *PAQ* is θ radians.

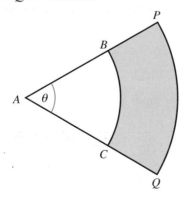

    a Find, in terms of θ, an expression for the perimeter of *BPQC*.

    b Given that the perimeter of *BPQC* is 14.5 cm, show that θ is 0.5.

    c Find, in cm², the area of *BPQC*.

9 In triangle ABC, $AB = 9$ cm, $AC = 6$ cm and angle $BAC = \frac{1}{6}\pi$ radians.
A circle, centre A and radius 2 cm, intersects the triangle at P and Q.
Calculate

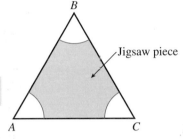

  a length BC,   **b** arc length QP,

  c the perimeter of the shaded region $QBCP$,   **d** the area of triangle ABC,

  e the area of sector AQP,   **f** the area of the shaded region $QBCP$.

10 A jigsaw piece is made from an equilateral triangle ABC with sides of length 2 cm.
A sector of a circle, radius 0.5 cm, is cut away from each vertex.

  a Find the perimeter of the jigsaw piece.

  b Find the area of the jigsaw piece.

Jigsaw piece

SKILLS CHECK **1B EXTRA** is on the CD

1.5 Trigonometric functions

Relate the periodicity and symmetries of the sine, cosine and tangent functions to the form of
their graphs.

Use the exact values of the sine, cosine and tangent of 30°, 45°, 60°.

You will need to be able to recall the main features of the graphs of
$y = \sin x$, $y = \cos x$ and $y = \tan x$. Make sure that you can sketch
them for values of x in degrees or radians.

Remember that $180° = \pi$ radians.

> **Note:**
> sin is shorthand for sine, cos for
> cosine and tan for tangent.

> **Tip:**
> In module C2 you may use a
> graphical calculator to check.

$y = \sin x$

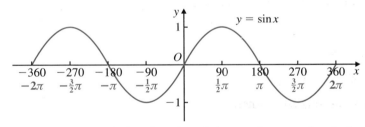

> **Note:**
> In degrees:
> $\sin(x + 360°) = \sin x$.
> In radians:
> $\sin(x + 2\pi) = \sin x$.

The minimum value of $\sin x$ is -1 and the maximum value is 1.
The graph is periodic, repeating every 360° (2π radians).
The vertical line through every vertex (turning point) is an axis of
symmetry.

$y = \cos x$

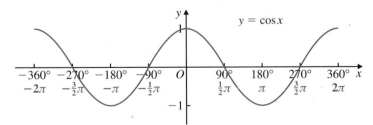

> **Note:**
> $y = \cos x$ is a translation of
> $y = \sin x$ by 90° ($\frac{1}{2}\pi$) to the left,
> i.e. $\cos x = \sin(x + 90°)$,
> or $\sin(x + \frac{1}{2}\pi)$ in radians.

The minimum value of $\cos x$ is -1 and the maximum value is 1.
The graph is periodic, repeating every 360^c (2π radians).
The vertical line through every vertex (turning point) is an axis of
symmetry, in particular the y-axis.

> **Note:**
> In degrees:
> $\cos(x + 360°) = \cos x$.
> In radians:
> $\cos(x + 2\pi) = \cos x$.

9

$y = \tan x$

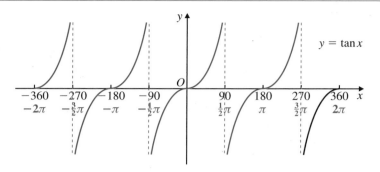

Note:
tan x takes every possible value in each 180° interval (π radians).

Notice that tan x can take any value.
The graph is periodic, repeating every 180° (π radians).
There are no lines of symmetry.

The graph has **asymptotes** at $x = \pm 90°$ $(\pm\frac{1}{2}\pi)$, $x = \pm 270°$ $(\pm\frac{3}{2}\pi)$, and so on.

Note:
In degrees:
$\tan(x + 180°) = \tan x$.
In radians:
$\tan(x + \pi) = \tan x$.

Note:
tan x is undefined at $\pm 90°$, $\pm 270°$, $\pm 450°$, …

Special angles

It is useful to recognise the sine, cosine and tangent of these special angles.

Note:
You must learn these.

x	$\sin x$	$\cos x$	$\tan x$
30° $(\frac{1}{6}\pi)$	$\dfrac{1}{2}$	$\dfrac{\sqrt{3}}{2}$	$\dfrac{1}{\sqrt{3}}$
45° $(\frac{1}{4}\pi)$	$\dfrac{1}{\sqrt{2}}$	$\dfrac{1}{\sqrt{2}}$	1
60° $(\frac{1}{3}\pi)$	$\dfrac{\sqrt{3}}{2}$	$\dfrac{1}{2}$	$\sqrt{3}$

1.6 Trigonometric identities

Use the identities $\tan \theta = \dfrac{\sin \theta}{\cos \theta}$ and $\sin^2 \theta + \cos^2 \theta = 1$.

The following **identities** are true for all values of θ:

$$\tan \theta = \frac{\sin \theta}{\cos \theta}$$

$$\sin^2 \theta + \cos^2 \theta = 1$$

Note:
In identities and equations, θ is often used for the angle.

Tip:
Learn these identities; you will often need to apply them when solving trigonometric equations.

Example 1.10 Find the exact value of tan θ, given that $\sin \theta = -\frac{5}{13}$ and $\cos \theta = \frac{12}{13}$.

Step 1: Use the appropriate trig identity and simplify if necessary.

$$\tan \theta = \frac{\sin \theta}{\cos \theta} = \frac{-\frac{5}{13}}{\frac{12}{13}} = -\frac{5}{12}$$

Tip:
Notice the instruction to give the *exact* value. You need to work in fractions here.

Example 1.11 Express $\dfrac{5 + 4\sin^2\theta}{3 - 2\cos\theta}$ in the form $a + b\cos\theta$, where a and b are integers to be found.

Step 1: Write each expression in terms of $\cos\theta$.

$$5 + 4\sin^2\theta = 5 + 4(1 - \cos^2\theta)$$
$$= 5 + 4 - 4\cos^2\theta$$
$$= 9 - 4\cos^2\theta$$
$$= (3 - 2\cos\theta)(3 + 2\cos\theta)$$

Step 2: Simplify the given expression.

$$\frac{5 + 4\sin^2\theta}{3 - 2\cos\theta} = \frac{(3 - 2\cos\theta)(3 + 2\cos\theta)}{3 - 2\cos\theta}$$
$$= 3 + 2\cos\theta$$

Step 3: Compare coefficients.

Comparing with $a + b\cos\theta$ gives $a = 3$ and $b = 2$.

> **Tip:**
> Rearrange $\sin^2\theta + \cos^2\theta = 1$ to get $\sin^2\theta = 1 - \cos^2\theta$.

> **Recall:**
> The difference between two squares:
> $p^2 - q^2 = (p - q)(p + q)$

1.7 Trigonometric equations

Find all the solutions, within a specified interval, of the equations $\sin(kx) = c$, $\cos(kx) = c$, $\tan(kx) = c$, and of equations (for example, a quadratic in $\sin x$) which are easily reducible to these forms.

The simplest trigonometric equations are of the form $\sin x = c$, $\cos x = c$ and $\tan x = c$, where c is a number.

Your calculator will give you *one* solution to an equation of this type, the **principal value** (PV). This lies in a particular range, depending on the function.

	In degrees	**In radians**
For sine function	$-90° \leqslant PV \leqslant 90°$	$-\tfrac{1}{2}\pi \leqslant PV \leqslant \tfrac{1}{2}\pi$
For cosine function	$0 \leqslant PV \leqslant 180°$	$0 \leqslant PV \leqslant \pi$
For tangent function	$-90° \leqslant PV \leqslant 90°$	$-\tfrac{1}{2}\pi \leqslant PV \leqslant \tfrac{1}{2}\pi$

You may be asked to find all the solutions in a given interval. To do this, find the principal value first. Then use the symmetries and periodicity of the graph to find further solutions in the the range.

You may be asked to give solutions in degrees or in radians. The following three examples are worked in degrees, with the answers in radians noted at the end of each part.

> **Tip:**
> To get the PV, key in $\sin^{-1}c$, $\cos^{-1}c$ or $\tan^{-1}c$ where c is a particular number. Make sure your calculator is in the correct mode: degrees or radians.

$\sin x = c$

Example 1.12 Find the values of x in the interval $0° \leqslant x \leqslant 360°$ for which

a $\sin x = 0.5$ **b** $\sin x = -0.9$

Step 1: Find the principal value PV.

Step 2: Use a sketch of $y = \sin x$ to find other values in the given range.

a From the calculator, $PV = \sin^{-1}(0.5) = 30°$.

The other value of x in the range is $180° - 30° = 150°$.

So $x = 30°, 150°$.

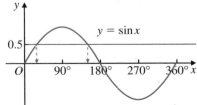

For x in radians, $PV = \tfrac{1}{6}\pi$.

In the interval $0 \leqslant x \leqslant 2\pi$, the solutions are $x = \tfrac{1}{6}\pi$ and $x = \pi - \tfrac{1}{6}\pi = \tfrac{5}{6}\pi$.

In decimal form, correct to two decimal places, $x = 0.52^c$ and $x = 2.62^c$.

> **Tip:**
> Key in
> $\boxed{\text{SHIFT}}\ \boxed{\text{SIN}}\ \boxed{0.5}\ \boxed{=}$

> **Tip:**
> Check on your calculator that $\sin x = 0.5$ for both values.

> **Tip:**
> Recognising this special angle (Section 1.5) enables you to give the solutions **exactly**, in terms of π.

b From the calculator, PV $= \sin^{-1}(-0.9) = -64.15...°$

Recall:
For the sin function, PV lies between $-90°$ and $90°$.

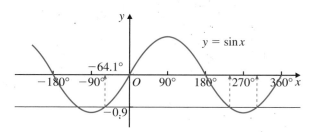

Other values of x in the interval $0° \leqslant x \leqslant 360°$ are

$180° + 64.15...° = 244.15...°$

$360° - 64.15...° = 295.84...°$

So $x = 244°, 296°$ (nearest degree).

Tip:
Check on your calculator, but remember that you rounded so you will not get $\sin x = -0.9$ exactly.

For x in radians, PV $= -1.12^c$. In the interval $0 \leqslant x \leqslant 2\pi$, the solutions are $x = \pi + 1.12^c$ and $x = 2\pi - 1.12^c$.

So $x = 4.26^c, 5.16^c$ (2 d.p.).

cos x = c

Example 1.13 Find the values of x in the interval $-360° < x < 360°$ for which

a $\cos x = \dfrac{\sqrt{3}}{2}$

b $\cos x = -0.8$

Step 1: Find the principal value PV.

a From the calculator, PV $= \cos^{-1}\left(\dfrac{\sqrt{3}}{2}\right) = 30°$.

Step 2: Use a sketch of $y = \cos x$ to find other values in the given range.

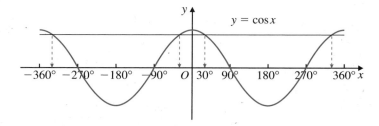

From the graph, other values of x are

$360° - 30° = 330°, -30°$ and

$-360° + 30° = -330°$.

So $x = -330°, -30°, 30°, 330°$.

The solutions can be written $x = \pm30°, \pm330°$.

In radians: PV $= \frac{1}{6}\pi$.

Tip:
Check these on your calculator.

Other solutions in the interval $-2\pi < x < 2\pi$ are $2\pi - \frac{1}{6}\pi, -\frac{1}{6}\pi$

and $-2\pi + \frac{1}{6}\pi$. So $x = \pm\frac{1}{6}\pi, \pm\frac{11}{6}\pi$.

Tip:
Since the y-axis is a line of symmetry, solutions will always be of the form $\pm$.

In decimal form, correct to two decimal places, $x = \pm0.52^c$ and $x = \pm5.76^c$.

b From the calculator, $PV = \cos^{-1}(-0.8) = 143.13\ldots°$

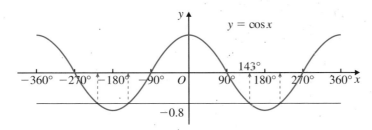

From the graph, other values of x are

$360° - 143.1\ldots° = 216.8\ldots°$,

$-143.13\ldots°$ and $-216.8\ldots°$.

So $x = \pm143°, \pm217°$ (nearest degree).

In radians: $PV = 2.5^c$ (1 d.p.).

Other solutions in the interval $-2\pi < x < 2\pi$ are $2\pi - 2.5^c$, -2.5^c and $-2\pi + 2.5^c$. So $x = \pm2.5^c, \pm3.8^c$.

Tip:
Use the fact that the curve is symmetrical about the y-axis.

$\tan x = c$

Example 1.14

Find the values of x in the interval $-180° \leqslant x \leqslant 180°$ for which $\tan^2 x = 3$.

Step 1: Form equations in the form $\tan x = c$.

$\tan^2 x = 3 \Rightarrow \tan x = \sqrt{3}$ or $\tan x = -\sqrt{3}$

Consider $\tan x = \sqrt{3}$.

Step 2: For each equation, find the PV.

From the calculator,
$PV = \tan^{-1}(\sqrt{3}) = 60°$.

Step 3: Use a sketch of $y = \tan x$ to find other values in the given range.

The other value of x in range is $-180° + 60° = -120°$.

So $x = -120°, 60°$.

Now consider $\tan x = -\sqrt{3}$.

From the calculator,
$PV = \tan^{-1}(-\sqrt{3}) = -60°$.

The other value of x in range is $180° - 60° = 120°$.

So $x = -60°, 120°$.

The complete solution is $x = \pm60°, \pm120°$.

Tip:
There are two equations hidden in this question.

Note:
For the tan function, when you have found a solution in an interval of $180°$, just add or subtract multiples of $180°$ to get further solutions.

In radians, where $-\pi < x < \pi$:

$\tan x = \sqrt{3}$ has $PV = \frac{1}{3}\pi$ and the other solution is $-\pi + \frac{1}{3}\pi = -\frac{2}{3}\pi$.

$\tan x = -\sqrt{3}$ has $PV = -\frac{1}{3}\pi$ and the other solution is $\pi - \frac{1}{3}\pi = \frac{2}{3}\pi$.

So the complete solution is $x = \pm\frac{1}{3}\pi, \pm\frac{2}{3}\pi$.

In decimal form, correct to two decimal places, $x = \pm1.05^c$ and $x = \pm2.09^c$.

Multiple angles

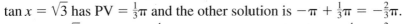

Take care if you are asked to solve equations involving multiples of x. This is illustrated in the following example.

Example 1.15   Find the values of x, in the interval $0° < x < 360°$, for which $\cos 2x = 0.5$.

Step 1: Make a substitution to get a simple equation.

Let $2x = \theta$, so the equation is $\cos \theta = 0.5$.

Step 2: Find the interval in which the new variable lies.

Interval required: $0° < x < 360°$

so $0° < 2x < 720° \Rightarrow 0° < \theta < 720°$

Step 3: Solve the equation in the new variable.

For θ, PV $= \cos^{-1}(0.5) = 60°$.

Other solutions in the interval $0° < \theta < 720°$ are

$360° - 60°, 360° + 60°, 720° - 60°$.

So $\theta = 60°, 300°, 420°, 660°$.

Step 4: Substitute back for x.

$2x = 60°, 300°, 420°, 660°$

$\Rightarrow x = 30°, 150°, 210°, 330°$

Using identities to solve trigonometric equations

Example 1.16   Find the values of θ, in the interval $-\pi < \theta < 2\pi$, for which $\sqrt{3} \sin \theta - \cos \theta = 0$.

Step 1: Rearrange to form an equation in $\tan \theta$, using an appropriate identity.

$$\sqrt{3} \sin \theta - \cos \theta = 0$$
$$\sqrt{3} \sin \theta = \cos \theta$$

$(\div \text{ by } \cos \theta)$ $\sqrt{3} \dfrac{\sin \theta}{\cos \theta} = 1$

$\sqrt{3} \tan \theta = 1$

$(\div \text{ by } \sqrt{3})$ $\tan \theta = \dfrac{1}{\sqrt{3}}$

Step 2: Find the PV.

PV $= \tan^{-1}\left(\dfrac{1}{\sqrt{3}}\right) = \frac{1}{6}\pi$

Step 3: Use the periodicity of the tan curve to find other solutions in the given interval.

Other solutions in range are

PV $+ \pi = \frac{1}{6}\pi + \pi = \frac{7}{6}\pi$

PV $- \pi = \frac{1}{6}\pi - \pi = -\frac{5}{6}\pi$

In the interval $-\pi < \theta < 2\pi$, $\theta = -\frac{5}{6}\pi, \frac{1}{6}\pi, \frac{7}{6}\pi$.

In decimal form, correct to two decimal places, $\theta = -2.62^c, 0.52^c, 3.67^c$.

Example 1.17   Find the values of θ in the interval $0 \leqslant \theta \leqslant 2\pi$ for which $2\sin^2 \theta = \sin \theta$, leaving your answers in terms of π.

Step 1: Form an equation $f(\theta) = 0$ and factorise if possible.

$2 \sin^2 \theta - \sin \theta = 0$

$\sin \theta (2 \sin \theta - 1) = 0$

Step 2: Solve the equations formed.

$\sin \theta = 0 \Rightarrow \theta = 0, \pi, 2\pi$

or $2 \sin \theta - 1 = 0 \Rightarrow \sin \theta = \frac{1}{2}$

$\theta = \frac{1}{6}\pi, \frac{5}{6}\pi$

So $\theta = 0, \frac{1}{6}\pi, \frac{5}{6}\pi, \pi, 2\pi$.

In decimal form, correct to two decimal places where appropriate,
$\theta = 0^c, 0.52^c, 2.62^c, 3.14^c, 6.28^c$.

Example 1.18   **a**   Write the expression $3 - 2\sin^2\theta - 3\cos\theta$ in terms of $\cos\theta$.

           b   Hence solve $3 - 2\sin^2\theta - 3\cos\theta = 0$ for values of θ in the interval $0 \le \theta \le 2\pi$.

Step 1: Using an appropriate identity, form an equation in $\cos\theta$ only.

a
$$3 - 2\sin^2\theta - 3\cos\theta = 3 - 2(1 - \cos^2\theta) - 3\cos\theta$$
$$= 3 - 2 + 2\cos^2\theta - 3\cos\theta$$
$$= 2\cos^2\theta - 3\cos\theta + 1$$

Tip:
Use $\sin^2\theta = 1 - \cos^2\theta$.

Step 2: Solve the equation in $\cos\theta$.

b
$$2\cos^2\theta - 3\cos\theta + 1 = 0$$
$$(\cos\theta - 1)(2\cos\theta - 1) = 0$$
$$\Rightarrow \qquad \cos\theta - 1 = 0$$
$$\cos\theta = 1$$
$$\theta = 0, 2\pi$$

or
$$2\cos\theta - 1 = 0$$
$$\cos\theta = \tfrac{1}{2}$$
$$\theta = \tfrac{1}{3}\pi, 2\pi - \tfrac{1}{3}\pi$$
$$= \tfrac{1}{3}\pi, \tfrac{5}{3}\pi$$

Tip:
If the expression does not factorise, use the quadratic formula.

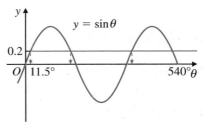

So, in the interval $0 \le \theta \le 2\pi$, $\theta = 0, \tfrac{1}{3}\pi, \tfrac{5}{3}\pi, 2\pi$.

In decimal form, correct to two decimal places where appropriate, $\theta = 0^c, 1.05^c, 5.24^c, 6.28^c$.

Example 1.19   Find the values of x in the interval $0° < x < 540°$ for which $\sin(x - 20°) = 0.2$. Give your answers correct to the nearest degree.

Step 1: Make a substitution to get a simple equation.

Let $x - 20° = \theta$.
The equation becomes $\sin\theta = 0.2$.

Tip:
Work out the interval for $x - 20°$.

Step 2: Find the interval in which the new variable lies.

Required interval: $0° < x < 540°$
$(-20°)$ $-20° < x - 20° < 520°$
Substitute θ $-20° < \theta < 520°$

Step 3: Work out the solutions for the new variable.

To solve $\sin\theta = 0.2$ for $-20° < \theta < 520°$, first find the PV.
$PV = \sin^{-1}(0.2) = 11.5...°$

$y = \sin\theta$

Other values of θ are

$$180° - 11.5...° = 168.4...°$$
$$360° + 11.5...° = 371.5...°$$
$$540° - 11.5...° = 528.4...° \text{ (just out of range)}$$

So $\theta = 11.5...°, 168.4...°, 371.5...°$.

Tip:
To maintain accuracy, work with uncorrected values if possible.

Step 4: Write the solutions in terms of x.

$$x - 20° = 11.5...°, 168.4...°, 371.5...°$$
$$x = 31.5...°, 188.4...°, 391.5...°$$

The values of x are $32°, 188°, 392°$ (correct to the nearest degree).

Tip:
Remember to give your answers to the required accuracy and in the given range.

1 Solve the following equations for $0° \leqslant x < 360°$. If your answer is not exact, give it correct to the nearest degree.

    a $\sin x = 0.3$ **b** $\cos x = 0.5$ **c** $\tan x = -1.5$

    d $\cos 3x = \dfrac{\sqrt{3}}{2}$ **e** $\sin 2x = -0.5$ **f** $\tan(0.5x) = 1$

2 Solve the following equations for $0 \leqslant x \leqslant \pi$. You may give your answers exactly in terms of π, or in decimal form correct to two decimal places.

> **Hint:**
> You may wish to use the table of special values on page 10.

    a $\sin x = \dfrac{\sqrt{3}}{2}$ **b** $\cos x = -\dfrac{1}{\sqrt{2}}$ **c** $\tan x = -\sqrt{3}$

    d $\sin 2x = -0.5$ **e** $\cos 3x = 0.5$ **f** $\tan 4x = \dfrac{1}{\sqrt{3}}$

3 Find all the values of x in the interval $-2\pi < x < 2\pi$ for which $\cos x = 0.75$, giving your answers in radians, correct to two decimal places.

4 Find all the values of x in the interval $-180° \leqslant x \leqslant 180°$ for which $2\cos^2 x = \sqrt{3}\cos x$.

5 Find all the values of x in the interval $-360° \leqslant x \leqslant 360°$ for which $\sqrt{2}\sin^2 x - \sin x = 0$.

6 Show that $\dfrac{4 + \cos^2 \theta}{5 - \sin^2 \theta} = 1$ for all values of θ.

7 Find the exact value of $\tan \theta$ given that $\sin \theta = -\frac{4}{5}$ and $\cos \theta = -\frac{3}{5}$.

 8 **a** Given that $\sin 3x = \cos 3x$, write down the value of $\tan 3x$.

    b Hence find all the solutions of the equation $\sin 3x = \cos 3x$ in the interval $0 < x < \pi$.

9 Find all the values of x in the interval $0 \leqslant x \leqslant 2\pi$ for which $2\sin\left(x + \dfrac{\pi}{3}\right) = 1$ giving your answer to two decimal places.

 10 **a** Given that $2\sin^2 x = 1 - \cos x$, show that $2\cos^2 x - \cos x - 1 = 0$.

    b Hence find all the values of x in the interval $0° \leqslant x \leqslant 360°$ for which $2\sin^2 x = 1 - \cos x$.

    c Write down all the values of x in the interval $0° \leqslant x \leqslant 180°$ for which $2\sin^2 2x = 1 - \cos 2x$.

SKILLS CHECK **1C EXTRA** is on the CD

Examination practice   Trigonometry

1 In triangle ABC, $AC = 50$ m, angle $BCA = 118°$ and angle $ABC = 35°$.

    i Calculate the length of AB, giving your answer to the nearest metre.

    ii Calculate the area of triangle ABC.

2 In triangle PQR, angle $PQR = 150°$ and $PQ = 42$ cm. The area of the triangle is $630\,\text{cm}^2$.

Calculate

i length QR,

ii length PR, giving your answer to the nearest mm.

3 The diagram shows a sector of a circle with centre O and radius r cm. The arc AB subtends an angle of $53°$ at O.

i Express $53°$ in radians, correct to 3 decimal places.

ii Given that the length of the arc AB is 37 cm, find the value of r.

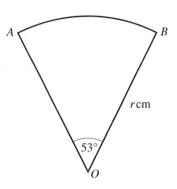

[OCR June 2002]

4 The diagram shows a sector of a circle with centre C and radius 20 cm. The angle ACB is θ radians. Given that the length of the arc AB is 46 cm, find

i the value of θ,

ii the area of the sector.

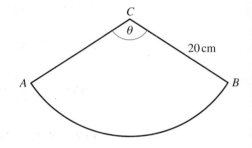

[OCR June 2001]

5 The diagram shows part of a circle with centre O and radius 15 cm. The major sector $OABC$ is shaded. The triangle AOC is equilateral. The reflex angle subtended at O by the major arc ABC is θ radians.

i Find the value of θ.

ii Find the perimeter of the major sector $OABC$, giving your answer correct to 3 significant figures.

iii Find the area of the major sector $OABC$, giving your answer correct to 3 significant figures.

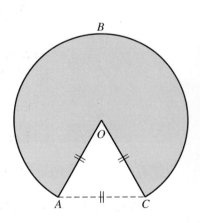

[OCR Nov 2002]

6 Find the exact value of $\cos 30° + 2 \sin 60°$.

[OCR June 2001]

7 Solve the equation $\sin \theta° = -\dfrac{\sqrt{3}}{2}$, giving all values of θ such that $0 \leqslant \theta \leqslant 360$.

[OCR Jan 2002]

8 a

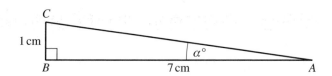

In triangle ABC shown in the diagram, $AB = 7$ cm, $BC = 1$ cm, angle $ABC = 90°$ and angle $BAC = \alpha°$. Find the value of $\cos \alpha°$, giving your answer in simplified surd form with a rational denominator.

b i Show that

$$3 \sin^2 \theta° + \cos^2 \theta° - \cos \theta° \tan \theta° - 1$$

may be written as $2 \sin^2 \theta° - \sin \theta°$.

ii Hence solve the equation

$$3 \sin^2 \theta° + \cos^2 \theta° - \cos \theta° \tan \theta° - 1 = 0,$$

giving all values of $\theta°$ such that $0 \leqslant \theta \leqslant 360$. [OCR Nov 2002]

9 i Express $3 \cos^2 \theta° - 2 \sin \theta°$ in terms of $\sin \theta°$.

ii Hence solve the equation

$$3 \cos^2 \theta° - 2 \sin \theta° = 2,$$

giving all values such that $0 \leqslant \theta \leqslant 360$. Where appropriate, give your answers correct to one decimal place. [OCR May 2002]

10 i Show that the equation $3 \sin 2\theta - \cos 2\theta = 0$ may be written as $\tan 2\theta = \frac{1}{3}$.

ii Hence solve the equation $3 \sin 2\theta - \cos 2\theta = 0$, giving all values of θ such that $0° \leqslant \theta \leqslant 360°$. Give your answers correct to the nearest $0.1°$. [OCR Jan 2001]

 11 Solve the equation $\cos (x + 0.6) = -0.5$, in the interval $0 < x < 2\pi$, giving your answers in radians, correct to two decimal places.

12 Solve the equation $\tan x = -\sqrt{3}$ in the interval $0° < x < 360°$.

 13 i Express $2 \cos^2 \theta - \sin^2 \theta - \sin \theta$ in the form $a \sin^2 \theta + b \sin \theta + c$, stating the values of the constants a, b and c.

ii Hence solve the equation

$$2 \cos^2 \theta - \sin^2 \theta - \sin \theta = 2,$$

giving all values of θ such that $0° \leqslant \theta \leqslant 360°$. Where appropriate, give your answers correct to the nearest $0.1°$. [OCR June 2001]

2 Sequences and series

Understand the idea of a sequence of terms and use definitions such as $u_n = n^2$, and relations such as $u_{n+1} = 2u_n$, to calculate successive terms and deduce simple properties. Understand and use Σ notation.

A **sequence** is a succession of terms that follow a rule.

One way of defining a sequence is to give the formula for the nth term. The sequence is generated by substituting positive integer values of n into the formula.

> **Note:**
> The nth term is often denoted by u_n, x_n or t_n.

Example 2.1 The nth term of a sequence is u_n, where $u_n = 2n - 3$ for $n \geqslant 1$. Write down the first four terms of the sequence.

Step 1: Substitute $n = 1, 2, 3$ and 4 into the formula.

$n = 1$	$u_1 = (2 \times 1) - 3 = -1$
$n = 2$	$u_2 = (2 \times 2) - 3 = 1$
$n = 3$	$u_3 = (2 \times 3) - 3 = 3$
$n = 4$	$u_4 = (2 \times 4) - 3 = 5$

The first four terms of the sequence are $-1, 1, 3, 5$.

> **Note:**
> This is an example of an arithmetic sequence. See Section 2.2.

Example 2.2 Find the first five terms of the sequence defined by
$u_n = 3 + \left(\frac{1}{2}\right)^n$, $n \geqslant 0$.

Step 1: Substitute $n = 0, 1, 2, 3$ and 4 into the formula.

$n = 0$	$u_0 = 3 + \left(\frac{1}{2}\right)^0 = 3$
$n = 1$	$u_1 = 3 + \left(\frac{1}{2}\right)^1 = 3.5$
$n = 2$	$u_2 = 3 + \left(\frac{1}{2}\right)^2 = 3.25$
$n = 3$	$u_3 = 3 + \left(\frac{1}{2}\right)^3 = 3.125$
$n = 4$	$u_4 = 3 + \left(\frac{1}{2}\right)^4 = 3.0625$

The sequence is $3, 3.5, 3.25, 3.125, 3.0625, \ldots$

> **Note:**
> In this example the first term is u_0.

Another way of defining a sequence is to give a formula $x_{n+1} = f(x_n)$, where a subsequent term is given in relation to the previous term. To generate the sequence, you must know the first term.

> **Note:**
> This is known as an **iterative formula**, or a **recurrence relation**.

Example 2.3 A sequence is defined by $x_{n+1} = 3x_n - 2$, with $x_1 = 4$. Calculate x_2, x_3 and x_4.

Step 1: Substitute $n = 1, 2$ and 3 into the formula.

$n = 1$	$x_2 = 3x_1 - 2 = (3 \times 4) - 2 = 10$
$n = 2$	$x_3 = 3x_2 - 2 = (3 \times 10) - 2 = 28$
$n = 3$	$x_4 = 3x_3 - 2 = (3 \times 28) - 2 = 82$

> **Note:**
> In this example the first term is x_1.

Calculator note:
If your calculator has an ANS key, try generating the terms as follows:

Display

Enter the first value   4   =   4

Write the formula   3 × ANS − 2 =   10

Generate next term   =   28

Generate next term   =   82

Tip:
You may not have to press ×.

Tip:
The next term is generated every time you press =. Keep on pressing!

Example 2.4   The value of a property is estimated to increase each year, where its price in each subsequent year is related to the current year by the relation $x_{n+1} = 1.05x_n$.

Given that the value of the property at the end of 2003 was £120 000, find the estimated value of the property at the end of 2007, to the nearest £1000.

Note:
This recurrence relation is describing a 5% increase in value each year.

Step 1: Substitute $n = 1, 2, 3, \ldots$ successively into the formula.

Value in 2003 $x_1 = 120\,000$

Value in 2004 $x_2 = 1.05x_1 = 1.05 \times 120\,000 = 126\,000$

Value in 2005 $x_3 = 1.05x_2 = 1.05 \times 126\,000 = 132\,300$

Value in 2006 $x_4 = 1.05x_3 = 1.05 \times 132\,300 = 138\,915$

Value in 2007 $x_5 = 1.05x_4 = 1.05 \times 138\,915 = 145\,860.75$

Note:
This is an example of a geometric sequence. See Section 2.3.

The estimated value of the property at the end of 2007, to the nearest £1000, is £146 000.

Using Σ notation

When the terms of a sequence are added, a **series** is formed, for example $2 + 6 + 10 + 14 + 18 + 22$.

There is a shorthand way of writing the terms of a series using a general term and Σ notation.

Note:
Σ means 'the sum of' and is read 'sigma'.

For example, to evaluate $\displaystyle\sum_{r=1}^{5} (2r + 3)$, add the terms calculated by substituting $r = 1, 2, 3, 4$ and 5.

When $r = 1$, $2r + 3 = 5$; when $r = 2$, $2r + 3 = 7$; and so on.

So $\displaystyle\sum_{r=1}^{5} (2r + 3) = 5 + 7 + 9 + 11 + 13 = 45$.

Example 2.5   Evaluate $\displaystyle\sum_{r=1}^{4} u_r$ where

Note:
Letters other than r may be used, for example i.

a $u_r = 6 - 3r$

b $u_r = 3(2^r)$.

20

Step 1: Substitute
$r = 1, 2, 3, 4$.

a $\quad u_1 = 6 - (3 \times 1) = 3 \qquad\qquad u_2 = 6 - (3 \times 2) = 0$

$\quad u_3 = 6 - (3 \times 3) = -3 \qquad\quad u_4 = 6 - (3 \times 4) = -6$

Note:
This is an arithmetic series
(Section 2.2).

Step 2: Add the terms. $\qquad \displaystyle\sum_{r=1}^{4} u_r = 3 + 0 + -3 + -6 = -6$

Step 1: Substitute
$r = 1, 2, 3, 4$.

b $\quad u_1 = 3(2^1) = 6 \qquad\qquad\quad u_2 = 3(2^2) = 12$

$\quad u_3 = 3(2^3) = 24 \qquad\qquad\quad u_4 = 3(2^4) = 48$

Note:
This is a geometric series
(Section 2.3).

Step 2: Add the terms. $\qquad \displaystyle\sum_{r=1}^{4} u_r = 6 + 12 + 24 + 48 = 90$

SKILLS CHECK 2A: Introducing sequences and series

1 The nth term, u_n, of a sequence is $n^2 - 1$, $n \geqslant 1$. Find the first four terms of the sequence.

2 If $x_n = 2 - \left(\frac{2}{3}\right)^n$, $n \geqslant 0$, find x_0, x_1, x_2 and x_3.

3 Calculate the next four terms of the sequence defined by the relation

$\quad$ **a** $\quad x_{n+1} = 2x_n$, $x_1 = 2$ $\qquad\qquad\qquad$ **b** $\quad x_{n+1} = 2x_n + 1$, $x_1 = 3$

4 A car cost £10 000 when new. It depreciated in value by 10% each year.

$\quad$ **a** $\quad$ Write down a relation for the value of the car from one year to the next.

$\quad$ **b** $\quad$ Using this relation, find the value of the car five years from when it was bought, giving your answer to the nearest £100.

5 A sequence of numbers is defined by $v_{n+1} = \sqrt{v_n}$, $v_1 = 100$.

$\quad$ Write down the next seven terms of the sequence, giving your answer correct to four decimal places where appropriate.

 6 A sequence is defined by the relation $x_{n+1} = \dfrac{5 - 4x_n}{x_n}$, $x_1 = -6$.

$\quad$ Calculate x_2, x_3, x_4, x_5, x_6 and x_7.

7 Evaluate $\displaystyle\sum_{i=0}^{5} u_i$ where $\quad$ **a** $\quad u_i = 3 + 4i$ $\qquad$ **b** $\quad u_i = 0.5(2^i)$.

 8 Evaluate $\qquad\qquad$ **a** $\quad \displaystyle\sum_{r=1}^{3} (20 - 3r)$ $\qquad$ **b** $\quad \displaystyle\sum_{r=1}^{4} 2^{r-1}$.

9 A sequence is defined by $u_{r+2} = 2u_{r+1} - 3u_r - 1$, where $u_1 = 1$, $u_2 = 0$. Find u_3, u_4 and u_5.

10 Evaluate $\qquad\qquad$ **a** $\quad \displaystyle\sum_{r=1}^{5} (r^2 - r)$ $\qquad$ **b** $\quad \displaystyle\sum_{r=4}^{6} \frac{r}{r+1}$.

SKILLS CHECK 2A EXTRA is on the CD

Recognise arithmetic progressions; use the formulae for the nth term and for the sum of the first n terms to solve problems involving arithmetic progressions, including the formula $\frac{1}{2}n(n+1)$ for the sum of the first n natural numbers.

The sequence defined by $u_n = 4n + 1$ is 5, 9, 13, 17, 21, …

This is an example of an **arithmetic progression**, where each successive term is obtained from the previous one by *adding* a constant amount, called the **common difference**.

Adding the terms of an arithmetic progression gives an **arithmetic series**, for example $5 + 9 + 13 + 17 + 21 + \cdots$

> **Note:**
> A formula for this sequence of the form $x_{n+1} = f(x_n)$ is $x_{n+1} = x_n + 4$, with $x_1 = 5$.

General expression for u_n

In an arithmetic series, the first term, u_1, is usually denoted by a and the common difference is denoted by d.

The terms are as follows:

$$u_1 = a$$
$$u_2 = a + d$$
$$u_3 = (a + d) + d = a + 2d$$
$$u_4 = (a + 2d) + d = a + 3d$$

Continuing the pattern, $u_n = a + (n - 1)d$.

In the series $5 + 9 + 13 + 17 + 21 + \cdots$, $a = 5$ and $d = 4$,

so $u_n = 5 + (n - 1) \times 4$
$$= 5 + 4n - 4$$
$$= 4n + 1, \text{ as expected.}$$

> **Tip:**
> The formula for u_n will be given in the examination but it is useful to learn it.

> **Note:**
> Notice that the formula for u_n is linear and the coefficient of n gives the value of d.

Example 2.6 The third term in an arithmetic progression is 8 and the eighth term is 18. Find

a the first term and the common difference,

b the 23rd term,

c the nth term.

Step 1: Use the nth term to form two equations in a and d.

a
$$a + 2d = 8 \quad ①$$
$$a + 7d = 18 \quad ②$$

Step 2: Solve the equations.

$②-①$
$$5d = 10$$
$$d = 2$$

Substituting in ①: $a + 4 = 8$
$$a = 4$$

The first term is 4 and the common difference is 2.

> **Recall:**
> Solving simultaneous equations (C1 Section 2.7).

Step 3: Use the nth term formula.

b $u_{23} = a + 22d = 4 + (22 \times 2) = 48$

c nth term $= a + (n - 1)d$
$$= 4 + (n - 1) \times 2$$
$$= 4 + 2n - 2$$
$$= 2n + 2$$

Sum of first n terms, S_n

The sum of the first n terms of an arithmetic series, with first term a and common difference d, is written S_n and is given by

$$S_n = \frac{n}{2}[2a + (n-1)d]$$

Alternatively, if you know the first term and the last term, l, use

$$S_n = \frac{n}{2}(a+l)$$

Tip:
These formulae for S_n will be given in the examination.

Tip:
The last term, l, is the same as the nth term and can be expressed as $l = a + (n-1)d$.

Example 2.7   Find the sum of the first 20 terms of the arithmetic series $4 + 7 + 10 + 13 + \cdots$

Step 1: Define a, d and n. $a = 4$, $d = 3$, $n = 20$

Step 2: Use the formula for S_n.

$$S_n = \frac{n}{2}[2a + (n-1)d]$$

$$S_{20} = \frac{20}{2}(2 \times 4 + 19 \times 3) = 650$$

Example 2.8   The first term of an arithmetic progression is 6 and the eighth term is twice the third term. Find the sum of the first ten terms.

Step 1: Use the nth term formula to write u_8 and u_3 in terms of d.

$$a = 6, \ u_8 = 6 + 7d, \ u_3 = 6 + 2d$$

Step 2: Use the given relationship to find d.

$$u_8 = 2u_3 \Rightarrow 6 + 7d = 2(6 + 2d)$$
$$6 + 7d = 12 + 4d$$
$$d = 2$$

Step 3: Use the formula for S_n.

$$S_{10} = \frac{10}{2}(2 \times 6 + 9 \times 2) = 150$$

Example 2.9   Ben's aunt gave him money on his birthday every year from his 15th birthday to his 30th birthday.

She gave him £200 on his 15th birthday. How much did she give him in total if on each subsequent year she gave him

a   £100,

b   £100 more than on his previous birthday?

Tip:
In **a**, each term is the total amount Ben had been given on and before the particular birthday.

Tip:
Take care when working out n.

Step 1: Identify the type of sequence.

a   Ben's age: 15   16   17   18   ...   30

Total given: £200 £300 £400 £500   ...   ?

The totals form an **arithmetic sequence** with $a = 200$, $d = 100$, $n = 16$.

Step 2: Identify the term required and use an appropriate formula.

The total amount Ben had received by his 30th birthday is given by the 16th term of the sequence.

$$u_n = a + (n-1)d$$
$$u_{16} = 200 + (15 \times 100) = 1700$$

Ben had been given £1700.

Step 1: Identify the type of series.

b Ben's age:

15	16	17	18	...	30

Amount given each year: £200 £300 £400 £500 ... ?

So, by his 30th birthday, the total amount given to Ben is
$200 + 300 + 400 + \cdots +$ the amount given on his 30th birthday

Step 2: Identify the sum required and use an appropriate formula.

This is an **arithmetic series**, $a = 200$, $d = 100$, $n = 16$.

The total amount is the sum of the series.

$$S_{16} = \frac{16}{2}(2\times200 + 15\times100) = 15\,200$$

Ben had been given £15 200.

Tip:
In **b**, the total amount is found by adding the amounts given each year.

Using Σ notation to describe an arithmetic series

The *n*th term in an arithmetic progression can be written in the form $dn + c$, where d is the common difference and c is a constant. Hence an arithmetic series, with common difference d, can be written

$$\sum_{r=1}^{n} (dr + c).$$

For example $\sum_{r=1}^{5} (4r + 1) = 5 + 9 + 13 + 17 + 21 = 65$.

Recall:
Σ means 'the sum of' (see Section 2.1).

Note:
Σ notation can be used for other series, such as geometric series. See Section 2.3.

Note:
The series is arithmetic, with common difference 4.

Example 2.10 Evaluate $\sum_{r=1}^{14} (3r - 2)$.

Step 1: Substitute integer values of *r*, from the lower to the higher number.

$$\sum_{r=1}^{14} (3r - 2) = 1 + 4 + 7 + \cdots + 40$$

Step 2: Identify the series and use an appropriate formula.

This is an arithmetic series with $a = 1$, $d = 3$, $l = 40$, $n = 14$.

$$\sum_{r=1}^{14} (3r - 2) = S_{14} = \frac{14}{2}(1 + 40) = 287$$

Tip:
Work out the first few terms and the last term.

Tip:
Use $S_n = \frac{n}{2}(a + l)$.

Example 2.11 Show that $1 + 2 + 3 + \cdots + n = \frac{n(n + 1)}{2}$.

Step 1: Identify the type of series.

This is an arithmetic series with *n* terms, where $a = 1$, $d = 1$, $l = n$.

Step 2: Find S_n.

$$S_n = \frac{n}{2}(a + l) = \frac{n}{2}(1 + n) = \frac{n(n + 1)}{2}$$

Note:
You could use
$S_n = \frac{n}{2}[2a + (n - 1)d]$.

The result in Example 2.12 can be expressed as follows:

The **sum of the first *n* natural numbers** is given by

$$\sum_{r=1}^{n} r = \frac{n(n + 1)}{2}$$

Note:
Natural numbers are the counting numbers 1, 2, 3, 4, ...

Tip:
This formula will be given in the examination.

Example 2.12 Evaluate

a $\sum_{r=1}^{100} r$ **b** $\sum_{r=20}^{50} r$.

Step 1: Use the Σr formula.

a $\sum_{r=1}^{100} r = 1 + 2 + 3 + \cdots + 100 = \frac{100\times101}{2} = 5050$

Step 1: Use the Σr formula in two stages, subtracting unwanted terms.

b $\displaystyle\sum_{r=20}^{50} r = \sum_{r=1}^{50} r - \sum_{r=1}^{19} r$

$$= \frac{50 \times 51}{2} - \frac{19 \times 20}{2}$$

$$= 1275 - 190$$

$$= 1085$$

Tip:
The series is
$20 + 21 + \ldots + 50$.

Tip:
To use the Σr formula, the lowest value of r must be 1.

The following relationship can be used to sum any arithmetic series:

$$\sum_{r=1}^{n} (dr + c) = d\sum_{r=1}^{n} r + nc$$

Note:
$\displaystyle\sum_{r=1}^{n} c = nc$

Example 2.13 Evaluate $\displaystyle\sum_{r=1}^{24} (3r + 2)$.

Method 1

Step 1: Expand, using Σ notation.

$$\sum_{r=1}^{24} (3r + 2) = 3\sum_{r=1}^{24} r + 24 \times 2$$

Step 2: Apply the formula for Σr.

$$= 3 \times \frac{24 \times 25}{2} + 48$$

$$= 948$$

Tip:
$n = 24$.

Method 2

Step 1: Write out the first few terms and the last.

Alternatively, you can use the expanded series and the formula for the sum of an arithmetic series:

$$\sum_{r=1}^{24} (3r + 2) = 5 + 8 + 11 + \cdots + 74$$

Step 2: Identify the series.

This is an arithmetic series with $a = 5$, $d = 3$, $l = 74$, $n = 24$.

Step 3: Use the appropriate formula.

$$S_n = \frac{n}{2}(a + l)$$

$$= \frac{24}{2}(5 + 74)$$

$$= 948$$

Tip:
Substitute $r = 1, 2, 3, \ldots, 24$.

Tip:
Choose the method you prefer.

SKILLS CHECK **2B: Arithmetic progressions**

1 Find the common difference and the sum of the first 20 terms of the following:

    a $12 + 17 + 22 + \cdots$ **b** $-2 - 5 - 8 - \cdots$

2 Find the nth term and the sum of the first 200 terms of the arithmetic series $\frac{1}{2} + \frac{3}{2} + \frac{5}{2} + \frac{7}{2} + \cdots$

3 The first term of an arithmetic progression is 8 and the seventh term is 26. Find

    a the common difference,   **b** the nth term,   **c** the sum of the first 25 terms.

 4 The third, fourth and fifth terms of an arithmetic series are $(4 + x)$, $2x$ and $(8 - x)$ respectively.

    a Find the value of x.

    b Find the first term and the common difference.

    c Find the sum of the first 30 terms of the series.

5 The cost to a company of training a student is £1000 for the first student. The cost is then reduced by £50 for the second student, by a further £50 for the third student and so on, so that the cost for the second student is £950, for the third student is £900 and so on.

   a How much will it cost to train the 20th student?

   b How much will it cost to train 20 students?

6 Evaluate   **a** $\displaystyle\sum_{r=1}^{38} r$ **b** $\displaystyle\sum_{r=1}^{18} (7r - 1)$.

7 Evaluate $\displaystyle\sum_{r=12}^{20} (\tfrac{1}{2}r + 3)$.

8 Find n if $\displaystyle\sum_{r=1}^{2n} (4r - 1) = \sum_{r=1}^{n} (3r + 59)$.

9 The nth term of an arithmetic sequence is u_n, where $u_n = 6 + 2n$.

   a Find the values of u_1, u_2 and u_3.

   b Write down the common difference of the arithmetic sequence.

   c Find the value of n for which $u_n = 46$.

   d Evaluate $\displaystyle\sum_{n=1}^{20} u_n$.

10 Sharon borrowed £5625 on an interest-free loan. She paid back £25 at the end of the first month, then increased her payment by £50 in each subsequent month, paying £75 at the end of the second month, £125 at the end of the third month and so on.

   a How many months did she take to pay off the loan?

   b What was the amount of her final month's repayment?

SKILLS CHECK **2B EXTRA** is on the CD

2.3 Geometric progressions

Recognise geometric progressions; use the formulae for the nth term and for the sum of the first n terms to solve problems involving geometric progressions.

Use the condition $|r| < 1$ for the convergence of a geometric series, and the formula for the sum to infinity of a convergent geometric series.

The sequence defined by $u_n = 3(2^{n-1})$ is 3, 6, 12, 24, 48, …

This is an example of a **geometric progression**, where each term is obtained from the previous one by *multiplying* by a constant. The constant is called the **common ratio**.

Adding the terms of a geometric progression gives a **geometric series**, for example $3 + 6 + 12 + 24 + 48 + \cdots$

Note:
A formula for this geometric progression in the form $x_{n+1} = f(x_n)$ is $x_{n+1} = 2x_n$ with $x_1 = 3$.

Tip:
To find the common ratio, divide any term by the previous term.

General expression for u_n

In a geometric progression the first term, u_1, is usually denoted by a and the common ratio is denoted by r.

The terms are as follows:

$$u_1 = a$$
$$u_2 = ar$$
$$u_3 = ar^2$$
$$u_4 = ar^3$$

Continuing the pattern, $u_n = ar^{n-1}$.

Tip:
The formula for u_n will be given in the examination but it is useful to learn it.

Example 2.14 Find the eighth term of the geometric progression 3, 6, 12, …

Step 1: Define a and r. $\qquad a = 3, \ r = 2$

Step 2: Use the formula $\qquad u_n = 3(2^{n-1})$
for u_n. $\qquad \Rightarrow u_8 = 3(2^7) = 384$

Example 2.15 A geometric progression has second term 12 and fourth term 48. Find the seventh term, given that the first term is negative.

Step 1: Use the nth term $\quad u_2 = 12 \Rightarrow ar = 12 \qquad$ ①
formula to form two $\quad u_4 = 48 \Rightarrow ar^3 = 48 \qquad$ ②
equations in a and r.

Step 2: Solve the $\quad$ ② ÷ ①: $\qquad \dfrac{ar^3}{ar} = \dfrac{48}{12}$
equations.
$$r^2 = 4$$
$$r = \pm 2$$

Substituting in ①:

When $r = 2$, $2a = 12 \Rightarrow a = 6$. This is not applicable, since $a < 0$.

When $r = -2$, $-2a = 12 \Rightarrow a = -6$.

Step 3: Use u_n formula $\quad$ Since $a < 0$, $r = -2$ and $a = -6$.
with $n = 7$. $\qquad u_7 = ar^6 = (-6) \times (-2)^6 = -384$

Tip:
Test which value of r gives a negative value of a.

Example 2.16 In January 2001, an investor put £1000 into a savings account with a fixed interest rate of 4% per annum. Interest is added to the account on 31 December each year and no further capital is invested.

a By what factor is the amount in the account increased when interest is added?

b How much will be in the account, to the nearest £, when interest has been added on 31 December 2015?

Step 1: Calculate the **a** For an interest rate of 4%, the amount grows by $(1 + \frac{4}{100})$ each
multiplying factor. $\qquad$ year, that is, by a factor of 1.04.

Step 2: Apply the **b** Amount in account:
multiplying factor for the
appropriate number of
years.

January 2001	1000	
December 2001	1000×1.04	
December 2002	$1000 \times 1.04 \times 1.04 = 1000(1.04)^2$	
$\vdots$	$\vdots$	$\vdots$
December 2015	$1000(1.04)^{15}$	$= 1800.94\ldots$

The amount in the account on 31 December 2015 is £1801 (nearest £).

Note:
The amount in December 2015 is the 16th term in a geometric sequence, first term 1000 and common ratio 1.04.

Sum of the first n terms, S_n

The sum of the first n terms of a geometric series is

$$S_n = \frac{a(1 - r^n)}{1 - r}$$

Note:
This formula will be given in the examination.

Example 2.17 Calculate the sum of the first ten terms of the geometric series
$2 + 6 + 18 + 54 + \cdots$

Step 1: Define a, r and n. $\quad a = 2, r = 3, n = 10$

Step 2: Use the formula for the sum S_n.
$$S_{10} = \frac{2(1 - 3^{10})}{1 - 3}$$
$$= 59\,048$$

Tip:
You could use $S_n = \frac{a(r^n - 1)}{r - 1}$.

Sum to infinity of a convergent geometric series

If r lies between -1 and 1, that is $|r| < 1$, then as n gets larger, the terms in a geometric series get smaller. The sum of the series tends to a limiting value and the series is said to converge.

This is known as the **sum to infinity**, S_∞, where $S_\infty = \dfrac{a}{1 - r}$.

Note:
When $|r| < 1$, as $n \to \infty$, $ar^{n-1} \to 0$.

Example 2.18 The third term of a geometric series is $\frac{8}{3}$ and the sixth term is $\frac{64}{81}$.

Find   **a**   the sum of the first twenty terms,

         b   the sum to infinity.

Step 1: Use the nth term formula to write u_3 and u_6 in terms of a and r.

a $u_3 = \frac{8}{3} \Rightarrow ar^2 = \frac{8}{3}$ ①

     $u_6 = \frac{64}{81} \Rightarrow ar^5 = \frac{64}{81}$ ②

Step 2: Solve to find a and r.

$$② \div ① \quad \frac{ar^5}{ar^2} = \frac{\frac{64}{81}}{\frac{8}{3}}$$

$$\Rightarrow \quad\quad r^3 = \frac{8}{27}$$

$$r = \sqrt[3]{\frac{8}{27}} = \frac{2}{3}$$

Tip:
Work in fractions.

Substituting in ①:

$$a \left(\tfrac{2}{3}\right)^2 = \tfrac{8}{3}$$

$$a = \tfrac{8}{3} \div \left(\tfrac{2}{3}\right)^2 = 6$$

Step 3: Use the formula for S_n.

$$S_{20} = \frac{a(1 - r^{20})}{1 - r} = \frac{6\left(1 - \frac{2}{3}^{20}\right)}{1 - \frac{2}{3}} = 17.9945\ldots = 18 \text{ (2 s.f.)}$$

Step 1: Use the sum-to-infinity formula.

b   Since $r = \frac{2}{3}$, $|r| < 1$ and the series converges.

$$S_\infty = \frac{a}{1 - r} = \frac{6}{1 - \frac{2}{3}} = 18$$

Tip:
This is what you might expect, given the value for S_{20}.

Using Σ notation to describe a geometric series

A geometric series with common ratio r can be written in the form

$$\sum_{i=1}^{n} p(r^i), \text{ where } p \text{ and } r \text{ are numbers.}$$

Tip:
Look for a power of r in the general term.

For example, $\sum_{i=1}^{4} 5(2^i) = 10 + 20 + 40 + 80 = 150$.

This series has common ratio 2, as expected.

Note:
In this text, i has been used for the summation to avoid confusion with the common ratio r. Be careful – this may not always be the case!

Example 2.19 Evaluate $\sum_{i=1}^{9} 3(2^i)$.

Step 1: Expand the series and identify a, r and n.

$$\sum_{i=1}^{9} 3(2^i) = 3(2^1) + 3(2^2) + \cdots + 3(2^9)$$

This is a geometric series with $a = 3 \times 2 = 6$, $r = 2$ and $n = 9$, so S_9 is required.

Step 2: Apply the geometric series formula for S_n.

Using $$S_n = \frac{a(1 - r^n)}{1 - r}$$

$$\sum_{i=1}^{9} 3(2^i) = S_9 = \frac{6(1 - 2^9)}{1 - 2} = 3066$$

Example 2.20 Evaluate $\sum_{i=0}^{\infty} \left(\tfrac{1}{2}\right)^i$.

Tip:
The sum to infinity is required.

Step 1: Expand the series and identify a, r and n.

$$\sum_{i=0}^{\infty} \left(\tfrac{1}{2}\right)^i = \left(\tfrac{1}{2}\right)^0 + \left(\tfrac{1}{2}\right)^1 + \left(\tfrac{1}{2}\right)^2 + \cdots = 1 + \tfrac{1}{2} + \tfrac{1}{4} + \cdots$$

This is a geometric series with $a = 1$, $r = \tfrac{1}{2}$. S_∞ is required.

Step 2: Apply the geometric series formula for S_∞.

$$\sum_{i=0}^{\infty} \left(\tfrac{1}{2}\right)^i = \frac{a}{1 - r} = \frac{1}{1 - \tfrac{1}{2}} = 2$$

Tip:
Check that $|r| < 1$ is satisfied.

SKILLS CHECK **2C: Geometric progressions**

1 For the following geometric series, find

i the seventh term   **ii** the sum of the first seven terms   **iii** the sum to infinity (if possible)

a $2 + 10 + 50 + \cdots$ **b** $7 + \tfrac{7}{2} + \tfrac{7}{4} + \cdots$ **c** $1 - 2 + 4 - 8 + \cdots$

2 Find the sum of the first ten terms of the geometric series $-2 - 6 - 18 - \cdots$

3 A property was valued at £70 000 at the start of 2001. If the projected increase in value is 3% per year, find the projected value of the property at the start of 2020. Give your answer to the nearest £1000.

4 A geometric series has first term a and common ratio r, where $r > 0$. The third term is $\tfrac{5}{2}$ and the seventh term is $\tfrac{5}{512}$.

   a Find the values of a and r.   **b** Find the sum to infinity of the series.

5 The sum to infinity of a geometric series is $\tfrac{3}{4}$ and the sum of the first two terms is $\tfrac{2}{3}$. The common ratio of the series is negative.

   a Find the common ratio.

   b Find the *exact* difference between the sum of the first five terms and the sum to infinity.

6 Evaluate   **a** $\sum_{i=1}^{9} 2(4^i)$ **b** $\sum_{i=1}^{\infty} \left(\tfrac{3}{4}\right)^i$

7 Evaluate $\sum_{i=1}^{6} (7^i + 1)$

8 On her first birthday, Belinda is given £1. In each subsequent year, she is given double the amount that she received in the previous year, so that she receives £2 on her second birthday, £4 on her third birthday and so on.

   a How much does she receive on her tenth birthday?

   b How much, in total, has she received when she is 10?

9 A geometric series has first term 2 and second term $2\sqrt{2}$.

   a Find the seventh term.

   b The sum of the first five terms is $p + q\sqrt{2}$. Find the values of p and q.

10 Evaluate $\displaystyle\sum_{i=1}^{\infty} 2^{-i}$

SKILLS CHECK **2C EXTRA** is on the CD

2.4 Binomial expansions

Use the expansion of $(a + b)^n$, where n is a positive integer, including the recognition and use of the notations $\dbinom{n}{r}$ and $n!$

Notice the pattern in these **binomial expansions**:

$$(a + b)^0 = 1$$
$$(a + b)^1 = a + b$$
$$(a + b)^2 = a^2 + 2ab + b^2$$
$$(a + b)^3 = a^3 + 3a^2b + 3ab^2 + b^3$$
$$(a + b)^4 = a^4 + 4a^3b + 6a^2b^2 + 4ab^3 + b^4$$
$$(a + b)^5 = a^5 + 5a^4b + 10a^3b^2 + 10a^2b^3 + 5ab^4 + b^5$$

The coefficients of the terms can be found using **Pascal's triangle.** The first few lines are as follows:

```
              1
           1     1
        1     2     1
     1     3     3     1
  1     4     6     4     1
1     5    10    10     5     1
```

> **Note:**
> A binomial expansion contains two terms in the bracket.

> **Note:**
> In any specific expansion the powers of a descend and the powers of b ascend. In any ab term, the powers of a and b add up to the power of the bracket.

> **Tip:**
> $a = 1a^1$

> **Tip:**
> A number is formed by adding the two numbers immediately above it.
> $6 + 4$
> 10

Example 2.21 Expand $(3 + 2x)^4$ in ascending powers of x.

Step 1: Recall the appropriate line of Pascal's triangle.

The appropriate line of Pascal's triangle is   1   4   6   4   1.

Step 2: Compare with $(a + b)^n$ and expand.

$$(3 + 2x)^4 = 3^4 + 4(3^3)(2x) + 6(3^2)(2x)^2 + 4(3)(2x)^3 + (2x)^4$$
$$= 81 + 216x + 216x^2 + 96x^3 + 16x^4$$

Using $\dbinom{n}{r}$ to find the coefficients

For positive integer values of n, $\dbinom{n}{r} = {}^nC_r = \dfrac{n!}{r!(n-r)!}$. It can be calculated using factorial notation, for example

> **Note:**
> $n! = n(n-1)(n-2)$
> $\qquad \ldots \times 3 \times 2 \times 1,$
> $5! = 5 \times 4 \times 3 \times 2 \times 1$
> $\qquad = 120$

$$\binom{5}{2} = \frac{5!}{2!3!} = \frac{5\times4\times3\times2\times1}{2\times1\times3\times2\times1} = 10.$$

Alternatively use the $\boxed{nCr}$ button on your calculator and key in $\boxed{5}\ \boxed{nCr}\ \boxed{2}\ \boxed{=}$.

Tip:
Make sure that you know how to find $\boxed{nCr}$ on *your* calculator.

These expanded results are very useful:

$$\binom{n}{0} = 1 \qquad\qquad \binom{n}{1} = n \qquad\qquad \binom{n}{2} = \frac{n(n-1)}{2!}$$

$$\binom{n}{3} = \frac{n(n-1)(n-2)}{3!} \qquad \cdots \qquad \binom{n}{n} = 1$$

For example, $\binom{10}{3} = \frac{10\times9\times8}{3!} = \frac{720}{6} = 120.$

Binomial expansion formula for $(a + b)^n$

In general, for positive integer values of n,

$$(a + b)^n = a^n + \binom{n}{1}a^{n-1}b + \binom{n}{2}a^{n-2}b^2 + \cdots + \binom{n}{r}a^{n-r}b^r + \cdots + b^n$$

$$= a^n + na^{n-1}b + \frac{n(n-1)}{2!}a^{n-2}b^2 + \cdots + b^n$$

Note:
In the examination, this formula will be given in the booklet. 2! is written 1.2.

Example 2.22 Expand $(2x + 3y)^4$.

Step 1: Compare with $(a + b)^n$ and expand using the formula.

Comparing with $(a + b)^n$, $a = 2x$, $b = 3y$, $n = 4$:

$$(2x + 3y)^4 = (2x)^4 + \binom{4}{1}(2x)^3(3y) + \binom{4}{2}(2x)^2(3y)^2 + \binom{4}{3}(2x)(3y)^3 + (3y)^4$$

Step 2: Simplify.

$$= (2x)^4 + 4(2x)^3(3y) + 6(2x)^2(3y)^2 + 4(2x)(3y)^3 + (3y)^4$$

$$= 16x^4 + 96x^3y + 216x^2y^2 + 216xy^3 + 81y^4$$

Tip:
To find the coefficients, use ^4C_r on your calculator, use the expanded form or, easiest of all, recall the appropriate line of Pascal's triangle.

Example 2.23 Given that $(2 - y)^9 = A + By + Cy^2 + \cdots$, find A, B and C.

Step 1: Compare with $(a + b)^n$ and expand using the formula.

Comparing with $(a + b)^n$, $a = 2$, $b = (-y)$, $n = 9$.

$$(2 - y)^9 = 2^9 + \binom{9}{1}2^8(-y) + \binom{9}{2}2^7(-y)^2 + \cdots$$

Step 2: Simplify.

$$= 2^9 + 9(2^8)(-y) + 36(2^7)(-y)^2 + \cdots$$

$$= 512 - 2304y + 4608y^2 + \cdots$$

$A = 512$, $B = -2304$ and $C = 4608$.

Tip:
You need to calculate the first three terms of the series.

Tip:
$\binom{9}{2} = \frac{9\times8}{2\times1} = 36$

Example 2.24 Find the coefficient of x^4 in the expansion of $(2 + \frac{1}{2}x)^5$.

Step 1: Identify the term required and calculate the coefficient.

The term required is $\binom{5}{4}2^1(\frac{1}{2}x)^4 = 5 \times 2 \times \frac{1}{16}x^4 = \frac{5}{8}x^4$.

The coefficient of x^4 is $\frac{5}{8}$.

Tip:
Take care with $(\frac{1}{2}x)^4$. Remember to raise $\frac{1}{2}$ to the power 4 as well as x.

Binomial expansion formula for $(1 + x)^n$

If the first term in the bracket is 1, you can use a simplified version of the expansion formula.

For positive integer values of n

$$(1 + x)^n = 1 + \binom{n}{1}x + \binom{n}{2}x^2 + \binom{n}{3}x^3 + \cdots + \binom{n}{r}x^r + \cdots + x^n$$

$$= 1 + nx + \frac{n(n-1)}{2!}x^2 + \frac{n(n-1)(n-2)}{3!}x^3 + \cdots + x^n$$

Example 2.25   **a**   Expand $(1 - 2x)^8$ in ascending powers of x, as far as the term in x^3.

b   By letting $x = 0.01$, use the first four terms to find an approximate value for 0.98^8.

Step 1: Compare with the general formula for $(1 + x)^n$, expand and simplify.

a $(1 - 2x)^8 = 1 + 8(-2x) + \dfrac{8\times7}{2!}(-2x)^2 + \dfrac{8\times7\times6}{3!}(-2x)^3 + \cdots$

$\qquad\qquad\quad = 1 - 16x + 112x^2 - 448x^3 + \cdots$

Tip:
Replace n with 8 and 'x' with '$-2x$' in the formula for $(1 + x)^n$.

Step 2: Substitute the x-value and calculate.

b   When $x = 0.01$, $1 - 2x = 1 - 0.02 = 0.98$.

So $0.98^8 \approx 1 - 16(0.01) + 112(0.01)^2 - 448(0.01)^3$

$\qquad\quad = 0.850752$

$\qquad\quad = 0.8508$ (4 d.p.)

Note:
On the calculator $0.98^8 = 0.85076\ldots$ so the approximation is correct to four decimal places.

Example 2.26   In the expansion of $(1 + kx)^n$, where k and n are positive integers, the coefficient of x is 15 and the coefficient of x^2 is 90.

a   Show that $k = \dfrac{15}{n}$ and find n and k.

b   Hence find the term in x^3.

Step 1: Compare with the general formula for $(1 + x)^n$.
Step 2: Expand and simplify.

a $(1 + kx)^n = 1 + n(kx) + \dfrac{n(n-1)}{2!}(kx)^2 + \cdots$

$\qquad\qquad\quad = 1 + nkx + \dfrac{n(n-1)}{2!}k^2x^2 + \cdots$

Coefficient of x: $nk = 15 \Rightarrow k = \dfrac{15}{n}$ ①

Step 3: Compare the terms and solve for the unknowns.

Coefficient of x^2: $\dfrac{n(n-1)}{2!}k^2 = 90$ ②

Substituting for k from ①:

$$\frac{n(n-1)}{2}\left(\frac{15}{n}\right)^2 = 90$$

$$\frac{225n(n-1)}{2n^2} = 90$$

$$225(n-1) = 180n$$

$$225n - 225 = 180n$$

$$45n = 225$$

$$n = 5$$

Recall:
Simultaneous equations (Core 1 Section 2.7).

Tip:
Cancel n, then multiply both sides by $2n$.

Substituting in ①: $k = \frac{15}{5} = 3$, so $n = 5$, $k = 3$.

Step 1: Substitute values into the appropriate term.

b   Term in x^3: $\dbinom{5}{3}(3x)^3 = 10 \times 3^3 \times x^3 = 270x^3$

1 The polynomial f(x) is given by $(2 - 3x)^4$. Find the binomial expansion of f(x), simplifying your terms.

2 Simplifying your terms, find the first four terms in the expansion of $(1 + 4y)^7$ in ascending powers of y.

3   **a** Expand $(3 - 2x)^5$ in ascending powers of x up to the term in x^2.

  b Find the values of A, B and C, where $(5 + 2x)(3 - 2x)^5 = A + Bx + Cx^2 + \cdots$

4   **a** Expand $(1 + 2x)^6$ in ascending powers of x up to the term in x^3.

  b Using your expansion, find an approximation for $(1.02)^6$, correct to four decimal places. You must write down sufficient working to show how you obtained your answer.

5 Using the first four terms, in ascending powers of x, of the expansion of $(1 - 4x)^7$, find an approximate value for $(0.996)^7$, to a suitable degree of approximation. You must write down sufficient working to show how you obtained your answer.

6 Expand and simplify $\left(\sqrt{2} + \sqrt{3}\right)^4 - \left(\sqrt{2} - \sqrt{3}\right)^4$, leaving your answer in the form $a\sqrt{6}$, where a is a positive integer.

7 The coefficient of x^2 is $\frac{3}{8}$ in the expansion of $\left(1 + \dfrac{x}{n}\right)^n$. Find the value of n.

8 It is given that $(1 + kx)^n = 1 - 4x + 7x^2 + \cdots$

  a Find n and k.

  b Hence find the term in x^3.

9   **a** Expand $(1 + ax)^6$ in ascending powers of x up to and including the term in x^2.

  b In the expansion of $(1 + bx)(1 + ax)^6$, the coefficients of x and x^2 are 20 and 171 respectively. Find a and b, given that they are integers.

10   **a** Write down the first four terms, in ascending powers of x, in the expansion of $(1 - 3x)^5$.

  b Find the coefficient of x^3 in the expansion of $(1 + x)(1 - 3x)^5$.

SKILLS CHECK **2D EXTRA** is on the CD

Examination practice   Sequences and series

1 Find the sum of the first 400 positive integers.   [OCR June 2002]

2 The first term of an arithmetic progression is 14 and the 20th term is 25.4.

  i Find the common difference.

  ii Find the sum of the first 500 terms.   [OCR Jan 2002]

3 A geometric progression $u_1, u_2, u_3, \ldots$ is defined by

$$u_1 = 5, \quad u_{n+1} = 0.6u_n.$$

i Find u_4.

ii Find the sum to infinity of the terms of the geometric progression. [OCR Jan 2002]

4 A geometric progression has first term 30 and common ratio 0.8. Find

i the 20th term, giving your answer correct to 3 significant figures,

ii the sum of the first 20 terms, giving your answer correct to 3 significant figures,

iii the sum to infinity. [OCR June 2002]

5 The fifth term of an arithmetic progression is 14 and the thirteenth term is 26.
Find the sum of the first 60 terms. [OCR June 2004]

6 The first term of a geometric progression is 24 and the second term is 18.
Find the sum to infinity. [OCR June 2002]

7 The nth term of a sequence is ar^{n-1}, where a and r are constants. The first term is 3 and the second term is $-\frac{3}{4}$. Find the values of a and r.

Hence find the value of $\displaystyle\sum_{n=1}^{\infty} ar^{n-1}$. [OCR June 1996]

 8 Evaluate $\displaystyle\sum_{r=1}^{100} (3r - 1)$.

9 A fitness programme includes a daily number of step-ups. The number of step-ups scheduled for Day 1 is 20 and the number on each successive day is to be 3 more than the previous day (i.e. 23 on Day 2, 26 on Day 3).

The programme also includes a daily time to be spent jogging. The time, T_N minutes, to be spent jogging on Day N is given by the formula

$$T_N = 15 \times 1.05^{N-1}.$$

i Find the total number of step-ups scheduled to be completed during the first 30 days of the fitness programme.

ii Verify that the daily jogging time first exceeds 60 minutes on Day 30.

iii Find the total time to be spent jogging during the first 30 days of the fitness programme. [OCR Jan 2003]

10 i Expand $(2 + 3x)^4$ completely, simplifying the coefficients.

ii Hence find the coefficient of x^2 in the expansion of

$$(1 - \tfrac{1}{2}x)^2(2 + 3x)^4.$$ [OCR June 2001]

 11 i Expand $(1 + x)^7$ in ascending powers of x up to and including the term in x^3.

ii Hence find the coefficient of y^3 in the expansion of

$$(1 + 3y + y^2)^7.$$ [OCR June 2004]

3 Algebra

3.1 The factor theorem

Use the factor theorem.

For a polynomial f(x), the **factor theorem** states that if a is a root of the equation f(x) = 0, then ($x - a$) is a factor of f(x), and vice versa.

This can be written

$$\text{f}(a) = 0 \Leftrightarrow (x - a) \text{ is a factor of f}(x).$$

For example,

- if f(3) = 0, then ($x - 3$) is a factor of f(x)

- if f(−4) = 0, then ($x + 4$) is a factor of f(x)

- if ($x - 5$) is a factor of f(x), then f(5) = 0

- if ($x + 6$) is a factor of f(x), then f(−6) = 0.

Note:
The symbol ⇔ means that the statement is true when read from left to right, or from right to left.

The factor theorem can be used to identify factors of f(x) and hence solve f(x) = 0, as in the following example.

Example 3.1 It is given that f(x) = $x^3 + 4x^2 + x - 6$.

  a Find the value of f(1).

  b Use the factor theorem to write down a factor of f(x).

  c Express f(x) as a product of three linear factors.

  d Hence solve the equation $x^3 + 4x^2 + x - 6 = 0$.

Step 1: Substitute the value into f(x).

  a f(1) = $1^3 + (4 \times 1^2) + 1 - 6 = 1 + 4 + 1 - 6 = 0$

Step 1: Use the factor theorem.

  b f(1) = 0 ⇒ ($x - 1$) is a factor of f(x).

Step 1: Write down the linear factor, find and factorise the quadratic factor.

  c f(x) = ($x - 1$)($ax^2 + bx + c$)

    so $x^3 + 4x^2 + x - 6 \equiv (x - 1)(ax^2 + bx + c)$

Equate x^3 terms:	$x^3 = ax^3$	$\Rightarrow a = 1$
Equate constants:	$-6 = -c$	$\Rightarrow c = 6$
Equate x^2 terms:	$4x^2 = bx^2 - ax^2$	$\Rightarrow 4 = b - a \Rightarrow b = 5$

    Hence f(x) = ($x - 1$)($x^2 + 5x + 6$)
             = ($x - 1$)($x + 2$)($x + 3$)

Note:
You could use algebraic division (see Section 3.3).

Step 1: Solve f(x) = 0 by the usual methods.

  d f(x) = 0 ⇒ ($x - 1$)($x + 2$)($x + 3$) = 0

    So $x = 1$, $x = -2$, $x = -3$.

If, in the above example, you had not been given a hint to find the first factor, then a good strategy would be to try numbers that are factors of the constant term. In this case, try factors of −6: 1, −1, 2, −2, 3, −3, 6, −6.

3.2 The remainder theorem

Use the remainder theorem.

The **remainder theorem** states that if a polynomial $f(x)$ is divided by a linear term $(x - a)$, the remainder will be $f(a)$.

The remainder theorem is sometimes written

$$\frac{f(x)}{x - a} = g(x) + \frac{f(a)}{x - a} \quad \text{or} \quad f(x) = g(x)(x - a) + f(a)$$

where $g(x)$ is another polynomial of one degree less than $f(x)$.

Note:
The factor theorem is a special case of the remainder theorem, because if $(x - a)$ is a factor of $f(x)$, then $f(a) = 0$, indicating that $x = a$ is a root of $f(x) = 0$.

Example 3.2 Find the remainder when $f(x) = x^3 - 4x^2 - 5x + 7$ is divided by $(x + 2)$.

Step 1: Substitute $x = -2$ into the polynomial.

$f(-2) = (-2)^3 - 4(-2)^2 - 5(-2) + 7 = -8 - 16 + 10 + 7 = -7$
The remainder is -7.

Tip:
Be careful with signs.

3.3 Algebraic division

Carry out simple algebraic division.

Note:
Algebraic division has the same structure as the traditional long division with numbers.

Example 3.3 Divide $x^3 + 2x^2 - 3x + 7$ by $x - 2$.

The steps of this process are outlined on a PowerPoint slide in Example 3.3 which can be found on the CD.

$$
\begin{array}{r}
x^2 + 4x + 5 \\
x - 2 \overline{\smash{\big)}\ x^3 + 2x^2 - 3x + 7} \\
\underline{x^3 - 2x^2} \\
4x^2 - 3x \\
\underline{4x^2 - 8x} \\
5x + 7 \\
\underline{5x - 10} \\
+17
\end{array}
$$

$$(x^3 + 2x^2 - 3x + 7) \div (x - 2) = x^2 + 4x + 5 + \frac{17}{x - 2}$$

Note:
The remainder 17 can be confirmed by substituting $x = 2$ into the polynomial.
$f(2) = 8 + 8 - 6 + 7 = 17$

Example 3.4 Use the factor theorem to find a linear factor of $f(x)$ where $f(x) = 2x^3 - 5x^2 - x + 6$. Hence express $f(x)$ as a product of three linear factors.

Step 1: Find a value of x such that $f(x) = 0$.

Try $x = 1$: $f(1) = 2 - 5 - 1 + 6 = 2 \neq 0$. Hence $(x - 1)$ is not a factor.
Try $x = -1$: $f(-1) = -2 - 5 + 1 + 6 = 0$.
Hence $(x + 1)$ is a factor of $f(x)$.

Tip:
Try factors of 6: 1, -1, ...

Step 2: Find the quadratic factor.

Divide $f(x)$ by $(x + 1)$:

$$
\begin{array}{r}
2x^2 - 7x + 6 \\
x + 1 \overline{\smash{\big)}\ 2x^3 - 5x^2 - \ x + 6} \\
\underline{2x^3 + 2x^2} \\
-7x^2 - x \\
\underline{-7x^2 - 7x} \\
6x + 6 \\
\underline{6x + 6} \\
0
\end{array}
$$

Note:
Using algebraic division is an alternative to the method of identities described in Example 3.1.

Step 3: Factorise the quadratic factor. $\quad 2x^2 - 7x + 6 = (2x - 3)(x - 2)$

Step 4: State the three linear factors. $\quad$ Hence $f(x) = (x + 1)(2x - 3)(x - 2)$.

Example 3.5 When $f(x) = ax^3 + bx^2 + 4x - 6$ is divided by $(x - 1)$ the remainder is -1. When $f(x)$ is divided by $x - 3$ the remainder is 42.

Find the values of a and b.

Step 1: Use the remainder theorem twice. By the remainder theorem, $f(1) = -1$.

Hence $a(1)^3 + b(1)^2 + 4(1) - 6 = -1$
$$a + b = 1 \qquad \text{①}$$

Also by the remainder theorem, $f(3) = 42$.

Hence $a(3)^3 + b(3)^2 + 4(3) - 6 = 42$
$$27a + 9b = 36 \qquad \text{②}$$
$$(\div 9) \qquad 3a + b = 4 \qquad \text{③}$$
$$a + b = 1 \qquad \text{①}$$

Step 2: Solve the equations simultaneously.
$$\text{③} - \text{①} \qquad 2a = 3$$
$$a = 1.5$$

Substitute into ①
$$1.5 + b = 1$$
$$b = -0.5$$

So $a = 1.5$ and $b = -0.5$.

> **Tip:**
> Simplify the equations before attempting to solve.

SKILLS CHECK **3A: Factor and remainder theorems**

1 For the following polynomials, write down the roots of $f(x) = 0$.

   a $f(x) = (2x - 1)(x - 2)(3x + 1)$ **b** $f(x) = (x^2 - 4)(x^2 - 9)$

   c $f(x) = (4x - 1)(3x + 4)(x - 2)$ **d** $f(x) = (x + 3)(x - 1)^2$

2 Find the roots of the following, giving your answer in surd form if appropriate.

   a $x^3 - 7x + 6 = 0$ **b** $x^3 + x^2 - 10x - 6 = 0$ **c** $x^3 + 6x^2 + 11x + 6 = 0$

3 Find the remainder when the given polynomial is divided by the given linear term.

   a $x^3 - 3x^2 + 2x - 1$ divided by $(x - 2)$.

   b $x^3 + 7x^2 + 8x + 10$ divided by $(x + 1)$.

   c $x^3 + 3x^2 - 4$ divided by $(x - 1)$.

   Explain the significance of the result in part **c**.

4 Divide the given polynomial by the given linear term.

   a $x^3 - 6x^2 + 4x - 3$ divided by $(x - 3)$.

   b $x^3 + 6x^2 - x + 1$ divided by $(x + 1)$.

   c $2x^3 + x^2 - 4x + 5$ divided by $(x - 2)$.

5 Find the roots of the equation $x^3 - 6x^2 + x + 14 = 0$ given that 2 is a root. Give your answers in surd form if appropriate.

 6 Use the factor theorem to find a linear factor of $f(x)$ where $f(x) = x^3 + 3x^2 + 3x + 1$.
Hence express $f(x)$ as a product of three linear factors.

7 When the polynomial $f(x) = ax^2 + bx - 1$ is divided by $(x - 2)$ the remainder is 15.
When $f(x)$ is divided by $(x + 1)$ the remainder is –3.
Find the values of a and b.

8 When the polynomial $f(x) = ax^3 + bx^2 + 3x + 4$ is divided by $(x + 1)$ the remainder is -4.
When $f(x)$ is divided by $(x - 2)$ the remainder is 38.
Find the values of a and b.

SKILLS CHECK **3A EXTRA is on the CD**

3.4 Exponential curves

Sketch the graph of $y = a^x$, where $a > 0$, and understand how different values of a affect the shape of the graph.

The graph of $y = a^x$ is called an **exponential** curve. When $x = 0$, $y = a^0 = 1$, so the curve goes through $(0, 1)$.

Recall:
An exponent is another name for an index or a power (C1 Section 1.1).

The shape of the curve depends on the value of a.

When $a > 1$:

As $x \to \infty$, $y \to \infty$.

As $x \to -\infty$, $y \to 0$.

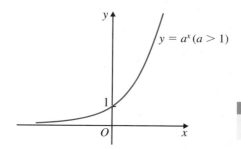

Note:
The y-value is always positive.

When $0 < a < 1$:

To get an idea of the general shape, let $a = \frac{1}{2}$ and consider $y = \left(\frac{1}{2}\right)^x$.

As $x \to \infty$, $y \to 0$.

As $x \to -\infty$, $y \to \infty$.

$\left(\frac{1}{2}\right)^x = 2^{-x}$, so when $a = \frac{1}{2}$ the curve is the graph of $y = 2^{-x}$. This is a reflection in the y-axis of $y = 2^x$.

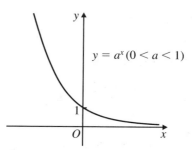

Note:
Try substituting very large and very small numbers for x.

3.5 Logarithms

Understand the relationship between logarithms and indices, and use the laws of logarithms.

The **logarithm** (log) of a positive number to a given **base** is the **power** to which the base must be raised to equal the number.

$$y = a^x \Leftrightarrow x = \log_a y$$

The base a is such that $a > 0$ and $a \neq 1$.

Note:
$\Leftrightarrow$ indicates a two-way implication: $y = a^x \Rightarrow x = \log_a y$ and $x = \log_a y \Rightarrow y = a^x$.

For example:

$$5^3 = 125 \iff \log_5 125 = 3 \quad \text{(5 is the base)}$$
$$10^2 = 100 \iff \log_{10} 100 = 2 \quad \text{(10 is the base)}$$

Example 3.6 Find x in each of the following.

a $\log_3 81 = x$ **b** $\log_3 1 = x$

c $\log_3 3 = x$ **d** $\log_3\left(\frac{1}{9}\right) = x$

Step 1: Write in index form.
Step 2: Solve the equation.

a $\log_3 81 = x \Rightarrow 3^x = 81$
By inspection $x = 4$, since $3^4 = 81$.

b $\log_3 1 = x \Rightarrow 3^x = 1$
By inspection $x = 0$, since $3^0 = 1$.

c $\log_3 3 = x \Rightarrow 3^x = 3$
By inspection $x = 1$, since $3^1 = 3$.

d $\log_3\left(\frac{1}{9}\right) = x \Rightarrow 3^x = \frac{1}{9}$
$$3^x = 3^{-2}$$
$$x = -2$$

Laws of logarithms

For any base a:
$$\log_a x + \log_a y = \log_a (xy)$$
$$\log_a x - \log_a y = \log_a\left(\frac{x}{y}\right)$$
$$k \log_a x = \log_a (x^k)$$

Examples:
$$\log_a 3 + \log_a 4 = \log_a 12$$
$$\log_a 20 - \log_a 5 = \log_a 4$$
$$2 \log_a 3 = \log_a 9$$

Special cases:
$$\log_a a = 1 \quad (\text{since } a^1 = a)$$
$$\log_a 1 = 0 \quad (\text{since } a^0 = 1)$$
$$\log_a (a^x) = x \log_a a = x$$
$$\log_a\left(\frac{1}{x}\right) = \log_a (x^{-1}) = -\log_a x$$

Examples:
$$\log_2 2 = 1$$
$$\log_3 1 = 0$$
$$\log_3 (3^4) = 4$$
$$\log_a\left(\tfrac{1}{3}\right) = -\log_a 3$$

Example 3.7 Find the value of x.

a $\log_a x = \log_a 20 - \log_a 15 + \log_a 3$

b $\log_a x = 2\log_a 8 - 3\log_a 4$

Step 1: Simplify using the log laws.

a $\log_a x = \log_a\left(\dfrac{20 \times 3}{15}\right) = \log_a 4$
$$\Rightarrow x = 4$$

b $\log_a x = 2\log_a 8 - 3\log_a 4$
$$= \log_a 8^2 - \log_a 4^3$$
$$= \log_a 64 - \log_a 64$$
$$= 0$$
$$\Rightarrow x = a^0 = 1$$

Example 3.8 Simplify $\log_a (a\sqrt{a})$.

$$\log_a(a\sqrt{a}) = \log_a a + \log_a (a^{\frac{1}{2}})$$
$$= 1 + \tfrac{1}{2} \log_a a$$
$$= 1 + \tfrac{1}{2}$$
$$= 1\tfrac{1}{2}$$

3.6 Exponential equations and inequalities

Use logarithms to solve equations of the form $a^x = b$, and similar inequalities.

One way of solving equations in the form $a^x = b$ is to take logs to the base 10 of both sides.

Example 3.9 Solve   **a** $5^x = 51$ **b** $2^{3x+1} = 12$

Step 1: Take logs to base 10 of both sides.

a $\log_{10}(5^x) = \log_{10} 51$

$x \log_{10} 5 = \log_{10} 51$

$x = \dfrac{\log_{10} 51}{\log_{10} 5}$

$= 2.44 \quad (3 \text{ s.f.})$

Step 2: Simplify using the log laws.

Step 3: Solve the equation in x.

b $\log_{10}(2^{3x+1}) = \log_{10} 12$

$(3x + 1) \log_{10} 2 = \log_{10} 12$

$3x + 1 = \dfrac{\log_{10} 12}{\log_{10} 2}$

$= 3.584 \ldots$

$3x = 3.584 \ldots - 1 = 2.584 \ldots$

$x = 0.862 \quad (3 \text{ s.f.})$

Note:
Use $\boxed{\log}$ on your calculator. This is programmed to give logs to base 10. You could also use $\boxed{\ln}$ which gives logs to the base e. You will learn more about $\log_e x$ in C3.

Tip:
These logs are divided, not subtracted, so do not try to cancel here.

You may be required to solve **inequalities** such as $a^x > b$. The initial stages of the working are similar to solving equations.

Example 3.10 Solve $3^x > 4000$.

Step 1: Take logs to base 10 of both sides.
Step 2: Simplify using the log laws.

$\log_{10}(3^x) > \log_{10} 4000$

$x \log_{10} 3 > \log_{10} 4000$

At this stage, you are recommended to find the log values from your calculator.

Step 3: Solve for x.

$0.477 \ldots x > 3.6020 \ldots$

$x > \dfrac{3.6020 \ldots}{0.477 \ldots}$

$x > 7.549 \ldots$

$x > 7.55 \quad (3 \text{ s.f.})$

Note:
$\log_{10} 3 = 0.477 \ldots$
$\log_{10} 4000 = 3.6020 \ldots$

A problem arises when negative numbers are involved, as in the following example.

Note:
If $0 < a < 1$, $\log a < 0$.

Example 3.11 Solve $0.3^x > 0.0025$.

Step 1: Take logs to base 10 of both sides.
Step 2: Simplify using the log laws.
Step 3: Solve for x.

$\log_{10}(0.3^x) > \log_{10} 0.0025$

$x \log_{10} 0.3 > \log_{10} 0.0025$

$-0.5228 \ldots x > -2.6020 \ldots$

$2.6020 \ldots > 0.5228 \ldots x$

$\dfrac{2.6020 \ldots}{0.5228 \ldots} > x$

$4.976 \ldots > x$

So $x < 4.98 \quad (3 \text{ s.f.})$

Note:
$\log_{10} 0.3 = -0.5228 \ldots$
$\log_{10} 0.0025 = -2.6020 \ldots$

Tip:
Avoid dividing by a negative quantity.

1 Evaluate   **a** $\log_4 64$ **b** $\log_5 25$ **c** $\log_2 8$ **d** $\log_{16} 4$.

2 Evaluate   **a** $\log_2 8^3$ **b** $\log_3 \left(\frac{1}{3}\right)$ **c** $\log_4 \sqrt{64}$ **d** $\dfrac{\log_3 27}{\log_3 9}$.

3 Find the value of x.

    a $\log_a x = \log_a 30 - \log_a 5 - \log_a 3$ **b** $\log_a x = 2\log_a 2 + 2\log_a 3$

4 Simplify   **a** $\log_a a^5$ **b** $\log_a \left(\dfrac{1}{\sqrt{a}}\right)$ **c** $4\log_a 1 + 3\log_a a$.

5 Express $\log_2 \sqrt{\dfrac{p^2 q}{2r^3}}$ in terms of $\log_2 p$, $\log_2 q$ and $\log_2 r$.

 6 Given that $\log_a x = 2(\log_a 3 + \log_a 2)$, where a is a positive constant, find x.

7 **a** Write down the value of   **i** $\log_3 3$ **ii** $\log_3 27$.

    b Find the value of $\log_3 2 - \log_3 54$.

8 Solve the following, giving your answers correct to three significant figures.

    a $2^x = 27$ **b** $3^{5x-2} \geqslant 20$ **c** $0.5^x < 0.05$

 9 **a** Show that $\log_a b + \log_a b^2 + \log_a b^3 + \cdots$ ($b \neq 1$) is an arithmetic series and state the common difference of the series.

    b The sum of the first ten terms of the series is $k \log_a b$. Find k.

10 It is given that $3^{2x} = 10(3^x) - 9$.

    a Writing $y = 3^x$, show that $y^2 - 10y + 9 = 0$.

    b Solve $3^{2x} = 10(3^x) - 9$.

SKILLS CHECK **3B EXTRA** is on the CD

Examination practice   Algebra

1 Given that $x - 2$ is a factor of $ax^3 + ax^2 + ax - 42$, find the value of the constant a.   [OCR Nov 2002]

2 The cubic polynomial $x^3 - 2x^2 - 2x + 4$ has a factor $(x - a)$, where a is an integer.

    i Use the factor theorem to find the value of a.

    ii Hence find exactly all three roots of the cubic equation
      $x^3 - 2x^2 - 2x + 4 = 0$.   [OCR June 1995]

3 Factorise $2x^3 - 3x^2 - 5x + 6$ completely. [OCR June 2003]

 4 i Express $2x^3 - 7x^2 + 2x + 3$ in the form $(x - 1)(ax^2 + bx + c)$, stating the values of a, b and c.

   ii Hence solve the equation $2x^3 - 7x^2 + 2x + 3 = 0$.

5 The polynomial f(x) is defined by

$$f(x) = x^3 + ax^2 - 2ax + c,$$

where a and c are constants.

   i It is given that $(x - 2)$ is a factor of f(x). Find the value of c.

   ii It is further given that, when f(x) is divided by $(x - 1)$, the remainder is 5. Find the value of a. [OCR Jan 2003]

6 The diagram shows the curve

$$y = -x^3 + 2x^2 + ax - 10.$$

The curve crosses the x-axis at $x = p$, $x = 2$ and $x = q$.

   i Show that $a = 5$.

   ii Find the exact values of p and q.

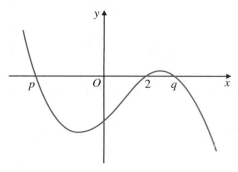

[OCR March 1996]

 7 The polynomial f(x) = $x^3 + ax^2 + bx + 5$ leaves a remainder of 1 when divided by $x + 1$ and leaves a remainder of 13 when divided by $x - 2$. Find the values of the constants a and b.

8 Express $\log_2(x + 2) - \log_2 x$ as a single logarithm.

Hence solve the equation $\log_2(x + 2) - \log_2 x = 3$. [OCR Nov 1995]

9 a Find the value of $\log_4 64$.

   b Given that $\log_x 27 = \log_4 64$, find the value of x.

 10 Find the value of x if $3^{x-2} = 2^{x+1}$, giving your answer correct to one decimal place.

11 i Given that $2 + \log_3 x = \log_3 y$, show that $y = 9x$.

   ii Hence, or otherwise, solve $2 + \log_3 x = \log_3(5x + 2)$.

12 i Express 4^{2x+1} as a power of 2.

   ii The variable x satisfies the equation $5^x \times 4^{2x+1} = 2^{x+3}$. Show that $x = \dfrac{\log_{10} 2}{\log_{10} 40}$.

13 Find the smallest integer satisfying the inequality $2^n > 10^{55}$.

4 Integration

Understand indefinite integration as the reverse process of differentiation.

The reverse process of differentiation is called **integration**.

This means that, given $\dfrac{dy}{dx}$, you integrate to find y; given $f'(x)$, you integrate to find $f(x)$.

Indefinite integration

You will have noticed when differentiating that more than one function can have the same derivative, for example:

$$y = 4x^2 \quad \Rightarrow \frac{dy}{dx} = 8x$$

$$y = 4x^2 - 6 \quad \Rightarrow \frac{dy}{dx} = 8x$$

In fact, for any constant c

$$y = 4x^2 + c \quad \Rightarrow \frac{dy}{dx} = 8x$$

When integrating, you take this into account by including a constant, called an **integration constant**. All possible values of the constant give a **family** of solutions, known as the **general solution**.

This process is called **indefinite integration**. There are several ways of writing it, including:

$$\frac{dy}{dx} = 8x \Rightarrow y = 4x^2 + c$$

$$f'(x) = 8x \Rightarrow f(x) = 4x^2 + c$$

$$\int 8x \, dx = 4x^2 + c$$

> **Recall:**
> If $y = ax^n$, $\dfrac{dy}{dx} = nax^{n-1}$
> (C1 Section 4.2).

> **Recall:**
> The derivative of a constant is zero (C1 Section 4.2).

> **Note:**
> Often c is used for the integration constant. It can take any value – positive, negative or zero.

> **Note:**
> $\int 8x \, dx$ is read 'the integral of $8x$, with respect to x'.

Integrate x^n (for any rational n except -1), together with constant multiples, sums and differences; solve problems involving the evaluation of a constant of integration.

To integrate x^n, reverse the differentiation process as follows:

- increase the power by 1

- divide by the new power.

For any rational number n, where $n \neq -1$,

$$\int x^n \, dx = \frac{1}{n+1}x^{n+1} + c$$

> **Note:**
> You can write
> $$\int x^n \, dx = \frac{x^{n+1}}{n+1} + c, n \neq -1.$$

For constant function k, $\int k\,dx = kx + c$.

If a is a constant

$$\int af(x)\,dx = a\int f(x)\,dx \Rightarrow \int ax^n\,dx = \frac{a}{n+1}x^{n+1} + c$$

Note:
You may prefer to write
$\frac{ax^{n+1}}{n+1} + c$.

Example 4.1 Find **a** $\int 3x^4\,dx$ **b** $\int 5t\sqrt{t}\,dt$.

Step 1: Write the term in index form if necessary.

a $\int 3x^4\,dx = \frac{3}{5}x^5 + c$

Note:
Letters other than x are often used.

Step 2: Integrate and simplify if necessary.

b $\int 5t\sqrt{t}\,dt = \int 5t^{\frac{3}{2}}\,dt$

Recall:
Laws of indices (C1 Section 1.1).

$$= \frac{5}{\frac{5}{2}}t^{\frac{5}{2}} + c$$

$$= 2t^{\frac{5}{2}} + c$$

Tip:
Take special care with fractions.

To integrate an expression in x containing several terms, such as $x^3 + 5x^2 + 2x - 1$, integrate term by term, using the rule

$$\int [f(x) \pm g(x)]dx = \int f(x)dx \pm \int g(x)dx$$

Example 4.2 Find $\int (x^3 + 5x)dx$.

Step 1: Integrate term by term.

$$\int (x^3 + 5x)dx = \frac{1}{4}x^4 + \frac{5}{2}x^2 + c$$

Tip:
Do not forget to include the integration constant c.

Example 4.3 Given that $\dfrac{dy}{dx} = 5x^4 - 6x + 3$, express y in terms of x.

Step 1: Integrate term by term.

$$y = \int (5x^4 - 6x + 3)dx$$

$$= \frac{5}{5}x^5 - \frac{6}{2}x^2 + 3x + c$$

Step 2: Simplify where possible.

$$= x^5 - 3x^2 + 3x + c$$

Tip:
This is asking you to integrate the function of x, where $y = \int \dfrac{dy}{dx}\,dx$.

Example 4.4 Find $\int \left(x + \dfrac{1}{x}\right)^2 dx$.

Step 1: Write the terms in the form ax^n using the index laws.

$$\left(x + \frac{1}{x}\right)^2 = \left(x + \frac{1}{x}\right)\left(x + \frac{1}{x}\right)$$

$$= x^2 + 1 + 1 + \frac{1}{x^2}$$

$$= x^2 + 2 + x^{-2}$$

Recall:
$\dfrac{1}{x^a} = x^{-a}$ (C1 Section 1.1).

$$\int \left(x + \frac{1}{x}\right)^2 dx = \int (x^2 + 2 + x^{-2})dx$$

Step 2: Integrate term by term.

$$= \frac{1}{3}x^3 + 2x + \frac{1}{-1}x^{-1} + c$$

$$= \frac{1}{3}x^3 + 2x - x^{-1} + c$$

Tip:
You could write this as
$\frac{1}{3}x^3 + 2x - \frac{1}{x} + c$

Example 4.5 **a** Write $\dfrac{4x^6 - x}{2x^4}$ in the form $ax^p + bx^q$ where a, b, p and q are rational numbers to be found.

b Hence find $\displaystyle\int \dfrac{4x^6 - x}{2x^4}\, dx$.

Step 1: Write the terms in the form ax^n using the index laws and compare with the given format.

a $\dfrac{4x^6 - x}{2x^4} = \dfrac{4x^6}{2x^4} - \dfrac{x}{2x^4} = 2x^2 - \tfrac{1}{2}x^{-3}$

Comparing coefficients gives

$a = 2,\ b = -\tfrac{1}{2},\ p = 2,\ q = -3.$

Step 2: Integrate term by term.

b $\displaystyle\int \dfrac{4x^6 - x}{2x^4}\, dx = \int (2x^2 - \tfrac{1}{2}x^{-3})\, dx$

$= \dfrac{2}{3}x^3 - \dfrac{\frac{1}{2}}{-2}x^{-2} + c$

$= \tfrac{2}{3}x^3 + \tfrac{1}{4}x^{-2} + c$

Sometimes you have additional information enabling you to find a specific value for the integration constant. In this case you can give a **particular solution**, rather than a general one, as in the following examples.

Example 4.6 A curve goes through the point $(1, 4)$ and $\dfrac{dy}{dx} = \dfrac{4}{x^3}$.

Find the equation of the curve.

Step 1: Write the term in the form ax^n using the index laws.

$\dfrac{dy}{dx} = \dfrac{4}{x^3} = 4x^{-3}$

Step 2: Integrate.

$y = \displaystyle\int 4x^{-3}\, dx$

$= \dfrac{4}{-2}x^{-2} + c$

$= -2x^{-2} + c$

Step 3: Substitute the coordinates of the given point to find c.

Since $(1, 4)$ lies on the curve, when $x = 1$, $y = 4$.

Substituting into the equation:

$4 = -2(1)^{-2} + c \Rightarrow c = 6$

The equation of the curve is $y = -2x^{-2} + 6$.

Example 4.7 Given that $f'(x) = \dfrac{(x + 2)^2}{3\sqrt{x}}$ and that $f(0) = 1$, find $f(x)$.

Step 1: Write terms in the form ax^n using the index laws.

$f'(x) = \dfrac{(x + 2)^2}{3\sqrt{x}} = \dfrac{x^2 + 4x + 4}{3x^{\frac{1}{2}}}$

$= \dfrac{x^2}{3x^{\frac{1}{2}}} + \dfrac{4x}{3x^{\frac{1}{2}}} + \dfrac{4}{3x^{\frac{1}{2}}}$

$= \tfrac{1}{3}x^{\frac{3}{2}} + \tfrac{4}{3}x^{\frac{1}{2}} + \tfrac{4}{3}x^{-\frac{1}{2}}$

Tip:
This is telling you to write the expression in index form before integrating.

Recall:
$\dfrac{x^m}{x^n} = x^{m-n}$ (C1 Section 1.1).

Tip:
Use your answer to part **a**.

Tip:
Be very careful with negatives and fractions. It is a good idea to write down the working.

Tip:
Remember to include the integration constant.

Tip:
You could write $y = 6 - \dfrac{2}{x^2}$.

Tip:
Expand the numerator and then divide each term in the numerator by $3x^{\frac{1}{2}}$.

Step 2: Integrate term by term.

$$f(x) = \int \left(\tfrac{1}{3}x^{\frac{3}{2}} + \tfrac{4}{3}x^{\frac{1}{2}} + \tfrac{4}{3}x^{-\frac{1}{2}}\right)dx$$

$$= \frac{\tfrac{1}{3}}{\tfrac{5}{2}}x^{\frac{5}{2}} + \frac{\tfrac{4}{3}}{\tfrac{3}{2}}x^{\frac{3}{2}} + \frac{\tfrac{4}{3}}{\tfrac{1}{2}}x^{\frac{1}{2}} + c$$

$$= \tfrac{2}{15}x^{\frac{5}{2}} + \tfrac{8}{9}x^{\frac{3}{2}} + \tfrac{8}{3}x^{\frac{1}{2}} + c$$

Step 3: Substitute the coordinates of the given point to find c.

Substituting $x = 0$: $f(0) = 0 + 0 + 0 + c = c$

$f(0) = 1 \Rightarrow c = 1$

$f(x) = \tfrac{2}{15}x^{\frac{5}{2}} + \tfrac{8}{9}x^{\frac{3}{2}} + \tfrac{8}{3}x^{\frac{1}{2}} + 1$

Tip:

$f(x) = \int f'(x)dx$

Recall:

$\dfrac{\tfrac{1}{3}}{\tfrac{5}{2}} = \tfrac{1}{3} \div \tfrac{5}{2} = \tfrac{1}{3} \times \tfrac{2}{5}$

SKILLS CHECK **4A: Integration methods**

1 Find

   a $\displaystyle\int 4x^2 \, dx$ **b** $\displaystyle\int \tfrac{2}{5} z^2 \, dz$ **c** $\displaystyle\int -6 \, dy$

2 Find y in terms of x if

   a $\dfrac{dy}{dx} = x^5 + 3x^2$ **b** $\dfrac{dy}{dx} = 2x^4 - \tfrac{1}{2}x^3$ **c** $\dfrac{dy}{dx} = x(2x + 7)$

   d $\dfrac{dy}{dx} = 4 - 3x$ **e** $\dfrac{dy}{dx} = -\tfrac{2}{5}x^3 - 1$ **f** $\dfrac{dy}{dx} = 2c$

 3 **a** Expand $(2x + 3)(x - 4)$.   **b** Find $\displaystyle\int (2x + 3)(x - 4)dx$.

4 **a** If $\dfrac{dA}{dt} = 4t^7$, find A in terms of t.

   b Given that $A = 0$ when $t = 1$, find the value of A when $t = 2$.

5 The gradient of a curve is given by $\dfrac{dy}{dx} = 5x - 3$. Find the equation of the curve, given that it passes through the origin.

6 Find

   a $\displaystyle\int \dfrac{3}{x^2} \, dx$ **b** $\displaystyle\int \dfrac{3}{\sqrt{x}} \, dx$ **c** $\displaystyle\int \sqrt[4]{t} \, dt$

7 **a** Express $x^2\sqrt{x}$ in the form x^k, where k is a rational number.   **b** Find $\displaystyle\int x^2\sqrt{x} \, dx$.

8 Find   **a** $\displaystyle\int (x^2 + \sqrt{x})dx$ **b** $\displaystyle\int \dfrac{x + 2}{\sqrt{x}} \, dx$.

9 The gradient of a curve is given by $\dfrac{dy}{dx} = \dfrac{1}{2\sqrt{x}}$. Find the equation of the curve, given that it passes through the point $(1, 3)$.

10 Given that $f'(x) = \tfrac{3}{2}\sqrt{x} - 1$ and $f(4) = 6$, find $f(x)$.

SKILLS CHECK **4A EXTRA is on the CD**

Evaluate definite integrals.

A **definite integral** has the form $\int_a^b f(x)dx$, where a and b are the **limits** of integration.

Note:
a is the lower limit and b is the upper limit.

To **evaluate** a definite integral:

- integrate $f(x)$ but omit the integration constant c
- substitute the upper limit
- subtract from this the value obtained when the lower limit is substituted.

Note:
You have to work out a specific value rather than give it as a function of x. It does not contain an integration constant c.

Example 4.8 Evaluate $\int_{-3}^{-2} (5 - 4x)dx$.

Step 1: Integrate the function.

$$\int_{-3}^{-2} (5 - 4x)dx = \left[5x - \tfrac{4}{2}x^2 \right]_{-3}^{-2}$$

Step 2: Substitute the limits, upper limit first.

$$= \left[5x - 2x^2 \right]_{-3}^{-2}$$

Step 3: Calculate the numerical value.

$$= 5(-2) - 2(-2)^2 - (5(-3) - 2(-3)^2)$$
$$= -10 - 8 - (-15 - 18)$$
$$= -18 - (-33)$$
$$= 15$$

Tip:
When you have integrated, use square brackets and put the limits on the right.

Note:
Take great care with the negative numbers.

Note:
The answer is a number, not a function of x.

Example 4.9 Evaluate $\int_1^\infty x^{-2}\, dx$

Step 1: Integrate.

$$\int_1^\infty x^{-2}\, dx = \left[\frac{1}{-1} x^{-1} \right]_1^\infty$$
$$= -\left[\frac{1}{x} \right]_1^\infty$$

Step 2: Consider the limits.

As $x \to \infty$, $\dfrac{1}{x} \to 0$, so

$$\int_1^\infty x^{-2}\, dx = -\left(0 - \frac{1}{1} \right)$$
$$= 1$$

Tip:
Consider the values of $\dfrac{1}{x}$ as x tends to zero.

4.4 The area under a curve

Use integration to find the area of a region bounded by a curve and lines parallel to the coordinate axes, or between two curves, or between a line and a curve.

Area under a curve

Integration can be used to find areas bounded by lines and curves. This is often referred to as finding the area 'under' a curve.

Consider the area of a region bounded by a curve $y = f(x)$, the x-axis and the lines $x = a$ and $x = b$.

If the area is *above* the x-axis:

$$\text{Area} = \int_a^b y\, dx$$

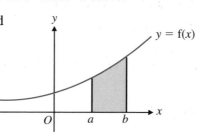

If the area is *below* the *x*-axis, the value of $\int_a^b y\, dx$ is negative.

The area is found by taking the positive value of the number calculated:

$$\text{Area} = \left| \int_a^b y\, dx \right|$$

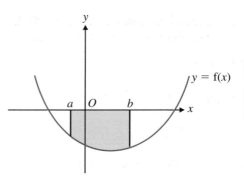

Note:
The straight lines mean that you take the positive value. This is called the modulus.

Example 4.10 Find the area of the region enclosed by the curve $y = x^2 + 2$, the *x*-axis and the lines $x = 1$ and $x = 2$.

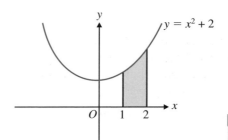

Step 1: Use the area formula with $a = 1$ and $b = 2$.

Step 2: Integrate term by term.

Step 3: Substitute the limits.

Step 4: Work out the values.

$$\int_a^b y\, dx = \int_1^2 (x^2 + 2)\, dx$$

$$= \left[\tfrac{1}{3}x^3 + 2x \right]_1^2$$

$$= [(\tfrac{1}{3} \times 2^3) + (2 \times 2)] - [(\tfrac{1}{3} \times 1^3) + (2 \times 1)]$$

$$= (\tfrac{8}{3} + 4) - (\tfrac{1}{3} + 2)$$

$$= 2\tfrac{2}{3} + 4 - 2\tfrac{1}{3}$$

$$= 4\tfrac{1}{3}$$

The area is $4\tfrac{1}{3}$.

Note:
The area is above the curve, so the integral will give a positive number.

Tip:
Be careful with the fractions. It is a good idea to write down all the working.

Example 4.11 The diagram shows a sketch of the curve $y = x(x - 3)$.

Find the area enclosed between the curve and the *x*-axis.

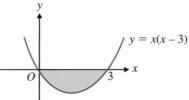

Step 1: Use the area formula with $a = 0$ and $b = 3$.

Step 2: Expand the brackets and integrate term by term.

Step 3: Substitute the limits and evaluate.

$$\int_a^b y\, dx = \int_0^3 x(x - 3)\, dx$$

$$= \int_0^3 (x^2 - 3x)\, dx$$

$$= \left[\tfrac{1}{3}x^3 - \tfrac{3}{2}x^2 \right]_0^3$$

$$= [(\tfrac{1}{3} \times 3^3) - (\tfrac{3}{2} \times 3^2)] - (0 - 0)$$

$$= 9 - 13\tfrac{1}{2}$$

$$= -4\tfrac{1}{2}$$

Step 4: Find the modulus.

$$\text{Area} = \left| \int_a^b y\, dx \right| = \left| -4\tfrac{1}{2} \right| = 4\tfrac{1}{2}$$

Tip:
Notice that the area is below the *x*-axis.

Tip:
Write the expression in index form before you integrate.

Tip:
It is helpful to put in the zeros to show that you have substituted for the lower limit.

Note:
As expected, the value of the integral is negative as the area lies below the *x*-axis.

To find the area between the curve $y = f(x)$, the y-axis and the lines $y = c$ and $y = d$, consider $\int_c^d x\,dy$.

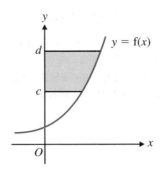

If the region is to the right of the y-axis.

$$\text{area} = \int_c^d x\,dy$$

If the region is to the left of the y-axis,

$$\text{area} = \left| \int_c^d x\,dy \right|$$

Example 4.12 The diagram shows the curve $y = x^2$
Calculate the area of the region enclosed between the curve, the y-axis and the lines $y = 4$ and $y = 9$.

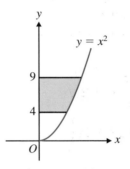

Step 1: Express x in terms of y.

$$y = x^2 \Rightarrow x = \sqrt{y} \quad (x > 0)$$

Step 2: Use the area formula with $a = 4$ and $b = 9$

$$\int_4^9 x\,dy = \int_4^9 \sqrt{y}\,dy$$

$$= \int_4^9 y^{\frac{1}{2}}\,dy$$

Step 3: Integrate.

$$= \left[\frac{1}{\frac{3}{2}} x^{\frac{3}{2}} \right]_4^9$$

Step 4: Substitute the limits and evaluate.

$$= \left[\tfrac{2}{3} x^{\frac{3}{2}} \right]_4^9$$

$$= \tfrac{2}{3}\left(9^{\frac{3}{2}} - 4^{\frac{3}{2}}\right)$$

$$= \tfrac{2}{3}(27 - 8)$$

$$= 12\tfrac{2}{3}$$

The area is $12\tfrac{2}{3}$.

> **Tip:**
> You can take out a factor of $\tfrac{2}{3}$.

Area enclosed between a line and a curve or between two curves

Let the upper line or curve be y_1, the lower line or curve be y_2 and the x-coordinates of the points of intersection be a and b ($a < b$).

Method 1: This involves finding the area under each line or curve and subtracting to find the enclosed area, where

$$\text{area} = \int_a^b y_1\,dx - \int_a^b y_2\,dx$$

Method 2: This involves finding an expression for $y_1 - y_2$ and integrating, where

$$\text{area} = \int_a^b (y_1 - y_2)\,dx$$

Example 4.13 The curve $y = 1 + 2\sqrt{x}$ and the line $y = x + 1$ intersect at $P(0, 1)$ and $Q(4, 5)$.

Find the area of the region enclosed between the line and the curve, shown shaded in the diagram.

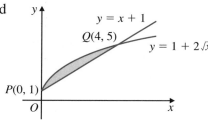

Method 1: Find the area under the curve and the line and subtract.

Step 1: Find the area 'under' the curve.

Let the area under the curve be A_1.

$$A_1 = \int_0^4 (1 + 2\sqrt{x})dx$$

$$= \int_0^4 (1 + 2x^{\frac{1}{2}})dx$$

$$= \left[x + \frac{2}{\frac{3}{2}}x^{\frac{3}{2}} \right]_0^4$$

$$= \left[x + \tfrac{4}{3}x^{\frac{3}{2}} \right]_0^4$$

$$= 4 + \tfrac{32}{3} - 0$$

$$= 14\tfrac{2}{3}$$

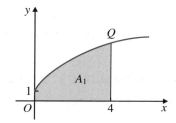

Step 2: Find the area 'under' the line.

Let the area under the line be A_2.

$$A_2 = \frac{1 + 5}{2} \times 4 = 12$$

Step 3: Subtract to find the required area.

Required area $= 14\tfrac{2}{3} - 12 = 2\tfrac{2}{3}$

Tip:
Area of a trapezium $= \tfrac{1}{2}$(sum of parallel sides) $\times$ perpendicular distance between them

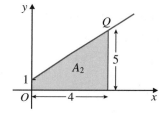

Method 2: Find the enclosed area by considering $y_1 - y_2$.

Step 1: Find $y_1 - y_2$.

Let $y_1 = 1 + 2\sqrt{x}$ and $y_2 = x + 1$.

$$y_1 - y_2 = 1 + 2\sqrt{x} - (x + 1) = 2\sqrt{x} - x$$

Area enclosed between the line and the curve:

Step 2: Integrate $y_1 - y_2$ with respect to x.

$$\int_a^b (y_1 - y_2)dx = \int_0^4 (2\sqrt{x} - x)dx$$

$$= \int_0^4 (2x^{\frac{1}{2}} - x)dx$$

$$= \left[\frac{2}{\frac{3}{2}}x^{\frac{3}{2}} - \tfrac{1}{2}x^2 \right]_0^4$$

$$= \left[\tfrac{4}{3}x^{\frac{3}{2}} - \tfrac{1}{2}x^2 \right]_0^4$$

$$= \tfrac{4}{3}(4)^{\frac{3}{2}} - \tfrac{1}{2}(4)^2 - 0$$

Step 3: Substitute the limits and evaluate.

$$= 2\tfrac{2}{3}$$

Note:
For $0 < x < 4$, $y_1 > y_2$, so $y_1 - y_2 > 0$.

Example 4.14 The diagram shows the curves $y = x^2 + 2$ and $y = 4 - x^2$.

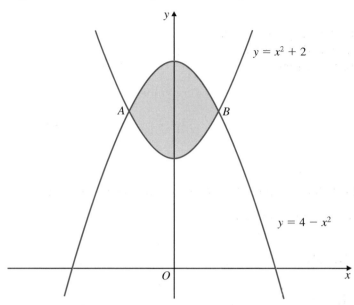

a Verify that the curves intersect at the points $A(-1, 3)$ and $B(1, 3)$.

b Show that the area of the shaded region between the curves is given by $\displaystyle\int_{-1}^{1} (2 - 2x^2)\,\mathrm{d}x$.

c Hence, or otherwise, show that the area of the shaded region between the curves is $2\frac{2}{3}$.

Step 1: Solve the equations simultaneously.

a When the curves intersect

$$x^2 + 2 = 4 - x^2$$

$$\Rightarrow \quad 2x^2 = 2$$

$$\Rightarrow \quad x^2 = 1$$

$$\Rightarrow \quad x = \pm 1$$

When $x = 1$, $y = 1^2 + 2 = 3$

When $x = -1$, $y = (-1)^2 + 2 = 3$

The curves intersect at $A(-1, 3)$ and $B(1, 3)$.

Note:
Alternatively, show that the coordinates of A and B satisfy both curves.

Step 2: Find $y_1 - y_2$.

b Let $y_1 = 4 - x^2$ and $y_2 = x^2 + 2$.

$$y_1 - y_2 = 4 - x^2 - (x^2 + 2) = 2 - 2x^2$$

Tip:
Make sure that $y_1 > y_2$, i.e. $y_1 - y_2 > 0$, between the points of intersection.

Step 3: Use the area formula.

$$\text{Area} = \int_a^b (y_1 - y_2)\,\mathrm{d}x$$

$$= \int_{-1}^{1} (2 - 2x^2)\,\mathrm{d}x$$

Step 4: Integrate.

c $\displaystyle \text{Area} = \left[2x - \tfrac{2}{3}x^3 \right]_{-1}^{1}$

Step 5: Substitute the limits and evaluate

$$= [2(1) - \tfrac{2}{3}(1)^3] - [2(-1) - \tfrac{2}{3}(-1)^3]$$

$$= 2 - \tfrac{2}{3} + 2 - \tfrac{2}{3}$$

$$= 2\tfrac{2}{3}$$

The area of the shaded region is $2\frac{2}{3}$.

1 Evaluate the following definite integrals:

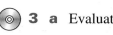

a $\displaystyle\int_1^2 3x^2\,dx$ **b** $\displaystyle\int_0^2 x^4\,dx$ **c** $\displaystyle\int_2^3 \tfrac{1}{2}x\,dx$ **d** $\displaystyle\int_2^\infty \frac{2}{x^2}\,dx$

2 a Evaluate $\displaystyle\int_{-2}^{2}(x+3)\,dx$ **b** Evaluate $\displaystyle\int_{-2}^{-1}(x^3+x)\,dx$

3 a Evaluate $\displaystyle\int_1^2\left(x-\frac{2}{x^2}\right)dx$ **b** Evaluate $\displaystyle\int_{-2}^{-1}\frac{1}{x^3}\,dx$

4 a Write $\dfrac{x^3+1}{x^2}$ in the form $x^p + x^q$, where p and q are integers.

 b Hence find the value of $\displaystyle\int_1^2\frac{x^3+1}{x^2}\,dx$

5 The diagram shows a sketch of the curve $y = 4x^3 + 1$.
Find the area of the region enclosed by the curve,
the x-axis and the lines $x = 1$ and $x = 2$.

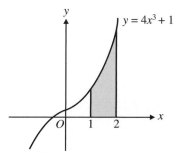

6 a Solve the simultaneous equations
$$y = x^2 + 2$$
$$x + y = 4$$

 b The curve $y = x^2 + 2$ and the line $x + y = 4$ intersect at
 P and Q.

 i Write down the coordinates of P and Q.

 ii Find the area enclosed between the line and the curve.

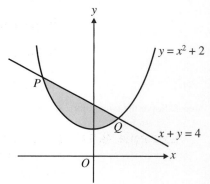

7 The diagram shows the graph of $y = -\dfrac{1}{x^3}$, for $x > 0$.

Find the area of the region enclosed between the curve,
the x-axis and the lines $x = 1$ and $x = 2$.

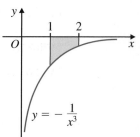

8 a Diagram 1 shows a sketch of the curve
$y = \sqrt[3]{x}$, for $x \geq 0$ and the line $y = 1$.

Show that the area enclosed by the curve,
the y-axis and the line $y = 1$ is 0.25.

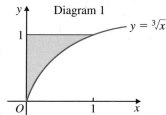

b Diagram 2 shows the same curve and
the line $y = x$.

Find the area of the region enclosed
between the line and the curve when $x \geq 0$.

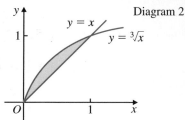

4.5 The trapezium rule

Use the trapezium rule to estimate the area under a curve, and use sketch graphs, in simple cases, to determine whether the trapezium rule gives an overestimate or an underestimate.

Consider the region enclosed by the curve $y = f(x)$, the x-axis and the lines $x = a$ and $x = b$. To find an approximation of the area of this region, split it into n intervals, of equal width h, where $h = \dfrac{b - a}{n}$. Form trapezia by joining the top ends of each interval with a straight line.

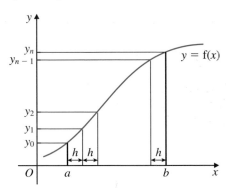

The area under the curve is given by $\displaystyle\int_a^b y \, dx$. An approximate value can be found using the **trapezium rule**:

$$\int_a^b y \, dx \approx \frac{1}{2} h \left[(y_0 + y_n) + 2(y_1 + y_2 + \cdots + y_{n-1}) \right] \text{ where } h = \frac{b - a}{n}.$$

Example 4.15 The diagram shows the region bounded by the curve $y = \sqrt{x - 1}$, the x-axis and the lines $x = 2$ and $x = 6$.

a Find an approximation to the area of the region, using the trapezium rule with four intervals.

b Is your value an overestimate or an underestimate?

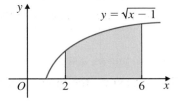

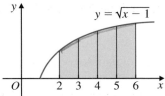

Step 1: Split the region into appropriate intervals.

a $h = \dfrac{6 - 2}{4} = 1$, $y_0 = \sqrt{2 - 1} = \sqrt{1}$, and so on.

x	2	3	4	5	6
y_n	y_0	y_1	y_2	y_3	y_4
$y = \sqrt{x - 1}$	$\sqrt{1}$	$\sqrt{2}$	$\sqrt{3}$	$\sqrt{4}$	$\sqrt{5}$

Step 2: Substitute values into the trapezium rule.

$$\int_a^b y \, dx \approx \frac{1}{2} \times 1[(\sqrt{1} + \sqrt{5}) + 2(\sqrt{2} + \sqrt{3} + \sqrt{4})]$$
$$= 6.764\ldots$$
$$= 6.76 \text{ (3 s.f.)}$$

Step 3: Decide whether the value is an overestimate or underestimate.

b From the graph, it can be seen that the value calculated by the trapezium rule is an underestimate for the area of the region.

Note on graphical calculators

Some graphical calculators can be programmed to find numerical approximations of definite integrals. These provide a useful check, but you will **not** be awarded marks for a question on the trapezium rule unless appropriate working has been shown.

Example 4.16 Find an approximation to $\int_0^1 \sqrt{\sin x}\,dx$, where x is in radians, using the trapezium rule with five intervals.

Step 1: Split the region into appropriate intervals.

$$h = \frac{1-0}{5} = 0.2$$

$y_0 = \sqrt{\sin 0} = 0$, $y_1 = \sqrt{\sin 0.2}$, ..., $y_5 = \sqrt{\sin 1}$.

x	0	1	2	3	4	5
y_n	y_0	y_1	y_2	y_3	y_4	y_5
$y = \sqrt{\sin x}$	0	$\sqrt{\sin 0.2}$	$\sqrt{\sin 0.4}$	$\sqrt{\sin 0.6}$	$\sqrt{\sin 0.8}$	$\sqrt{\sin 1}$

Step 2: Substitute values into the trapezium rule.

$$\int_0^1 \sqrt{\sin x}\,dx \approx \tfrac{1}{2} \times 0.2\,[(0 + \sqrt{\sin 1}) + 2(\sqrt{\sin 0.2} + \sqrt{\sin 0.4}$$
$$+ \sqrt{\sin 0.6} + \sqrt{\sin 0.8})]$$
$$= 0.6253\ldots$$
$$= 0.625 \text{ (3 s.f.)}$$

> **Tip:**
> Remember to set your calculator to radian mode.

SKILLS CHECK **4C: Trapezium rule**

Give answers to three significant figures unless requested otherwise.

1 a Sketch the graph of $y = 2^x$.

b Estimate $\int_0^4 2^x\,dx$, using the trapezium rule with four intervals.

c State whether your estimate is an overestimate or an underestimate.

2 The diagram shows a sketch of $y = \dfrac{1}{1+x}$ for $x > -1$.

Estimate $\int_1^2 \dfrac{1}{1+x}\,dx$ using the trapezium rule with five intervals.

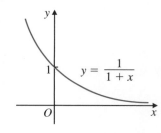

3 The following is a set of values, correct to three decimal places, for $y = \cos x$, where x is in radians.

x	0	0.2	0.4	0.6	0.8	1
y	1	0.980	0.921	p	0.697	q

a Find the value of p and the value of q.

b Use the trapezium rule and the values of y in the completed table to obtain an estimate for $\int_0^1 \cos x\,dx$, giving your answer to two decimal places

4 The diagram shows a sketch of $y = \log_{10} x$.

a Divide the shaded region into five equal-width intervals.

b Use the trapezium rule to estimate $\int_1^3 \log_{10} x\,dx$.

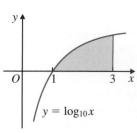

5 a Sketch the graph of $y = 10^{-x}$.

b Estimate $\int_{-2}^{-1} 10^{-x}\,dx$, using the trapezium rule with five intervals.

6 The diagram shows the graph of $xy = 12$, for $x > 0$.

Using the trapezium rule with six intervals, estimate the value of $\int_1^7 \dfrac{12}{x}\,dx$.

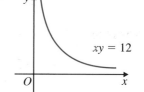

 7 The diagram shows a sketch of $y = 4 - x^2$.

a Estimate the area of the region enclosed between the curve and the x-axis, using the trapezium rule with four intervals.

b Evaluate the area exactly using integration and calculate the percentage error in taking your answer in **a** as the area.

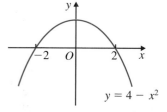

8 a Sketch the curve $y = 3^x + 1$, for values of x between -1 and 3.

b Use the trapezium rule with four intervals to estimate the area of the region bounded by the curve, the x-axis and the lines $x = -1$ and $x = 3$.

c Explain briefly how the trapezium rule could be used to find a more accurate estimate of the area of the required region.

9 The values of a function $f(x)$ are given in the table.

x	1	2	3	4	5	6
$f(x)$	3.5	5	7.5	11	15.5	21

Find an approximate value, using the trapezium rule with five intervals, for

a $\int_1^6 f(x)\,dx$

b $\int_1^6 \dfrac{1}{f(x)}\,dx$.

10 a Tabulate, correct to two decimal places, the values of the function $f(x) = \dfrac{2}{2 + x^2}$ for values of x from 0 to 2 at intervals of 0.4.

b Use the values found in part **a** to estimate $\int_0^2 \dfrac{2}{2 + x^2}\,dx$.

SKILLS CHECK **4C EXTRA** is on the CD

Examination practice   Integration

1 Find $\int \left(x^3 + 2x + \dfrac{1}{2x^2} \right) dx$.
         [OCR May 2002]

2 Find the equation of the curve through $(0, 5)$ for which $\dfrac{dy}{dx} = -2x + 3$.
       [OCR Jan 2002]

3 Find

  i $\int x(x + 1)\,dx$,
     ii $\int \dfrac{1}{x^2}\,dx$.
       [OCR Jan 2002]

4 a Find $\displaystyle\int_4^9 (2x - 3x^{\frac{1}{2}} + 1)\,dx$.

b Find $\displaystyle\int_2^\infty \frac{1}{x^3}\,dx$.

[OCR Jan 2004]

5 The gradient of a curve at the point (x, y) is given by $\dfrac{dy}{dx} = 3x^2 - x$. Use integration to find the equation of the curve, given that the curve passes through the point $(2, 1)$.

6 The gradient of a curve is given by

$$\frac{dy}{dx} = x^2(7\sqrt{x} - 3), \quad x > 0.$$

i Use integration to find y in terms of x.

ii Given that $y = 5$ at $x = 1$, find y in terms of x.

7 $f'(x) = k\sqrt{x} - 3x^2 + 41, \quad x > 0$.

i Given that $f'(4) = -1$, find the value of the constant k.

ii Given also that $f(4) = 134$, find $f(x)$.

8 a Find $\displaystyle\int (x^2 + 2)\,dx$.

b i Find $\displaystyle\int \frac{3}{x^2}\,dx$.

ii Evaluate $\displaystyle\int_1^\infty \frac{3}{x^2}\,dx$.

[OCR May 2003]

9 The diagram shows a sketch of the curve $y = 1 + \dfrac{1}{x^2}$ for $x > 0$.

The line $y = 2$ intersects the curve at $(1, 2)$.

Find the area of the region enclosed by the curve and the lines $x = 2$ and $y = 2$.

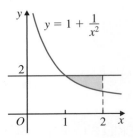

10 The diagram shows the curve $y = 3x^2 - 12x + 9$.

i Show that $\displaystyle\int_0^3 (3x^2 - 12x + 9)\,dx = 0$.

ii State what may be deduced from the result in part **i** about the areas labelled A and B.

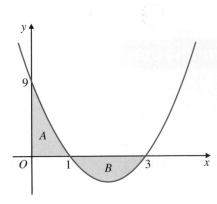

[OCR Jan 2002]

11 The diagram shows the curve $y = 8 - 2x^2$, together with the straight line $y = 5 - x$.

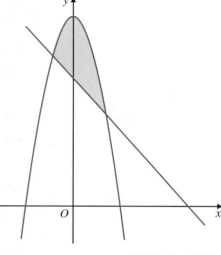

   i Find the coordinates of the points where the line and the curve intersect.

   ii Hence, or otherwise, solve the inequality $8 - 2x^2 < 5 - x$.

   iii Show that the area of the shaded region enclosed by the line and the curve is $\frac{125}{24}$.

[OCR Jan 2001]

12 The diagram shows the curves $y = x^2 + x + 5$ and $y = 2x^2 - 2x + 1$, and the shaded region R, between the curves. Determine the x-coordinate of each of the points of intersection of the two curves, and find the exact area of R.

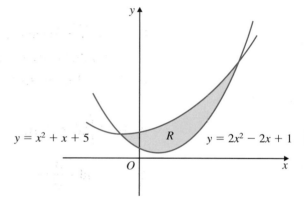

[OCR June 1999]

13 The diagram shows a sketch of $y = 2^{-x}$.

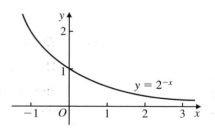

   i Use the trapezium rule with five intervals to find an approximation for $\int_0^2 2^{-x}\, dx$.

   ii By considering the graph of $y = 2^{-x}$, explain with the aid of a diagram whether your approximation will be an overestimate or an underestimate of the true value of $\int_0^2 2^{-x}\, dx$.

14 The diagram shows a sketch of $y = \dfrac{1}{1 + x^2}$.

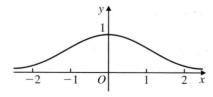

Use the trapezium rule with four intervals to find an approximation for $\displaystyle\int_{-2}^{2} \frac{1}{1 + x^2}\, dx$.

Practice exam paper

Answer **all** questions.

Time allowed: 1 hour 30 minutes

A calculator **may** be used in this paper.

1 The polynomial f(x) is defined by

f(x) = $2x^3 + x^2 - x + k$

where k is a constant. It is given that when f(x) is divided by ($x - 1$) the remainder is 12.

  i Show that $k = 10$. *(2 marks)*

  ii Hence express f(x) as the product of a linear factor and a quadratic factor. *(3 marks)*

2 Solve the equation $\log_4 x + \log_4(x + 12) = 3$. *(6 marks)*

3 Given that $\sin^2 x = 2\cos x$, find the exact value of $\cos x$. *(4 marks)*

Hence solve the equation $\sin^2 x - 2\cos x = 0$ for all values of x in the range $0° \leqslant x \leqslant 360°$. *(2 marks)*

4 A and B are two points on a circle which has centre O and radius 6 cm.
The acute angle AOB is θ radians, and the area of triangle ABO is 9 cm^2.

  i Calculate the exact value of θ. *(3 marks)*

  ii Calculate the area of the segment of the circle between the line AB and the minor arc AB. *(4 marks)*

5 In triangle PQR, $PQ = 2$ cm, $PR = 3$ cm, and angle $RPQ = 60°$. Calculate

  i the length QR, *(3 marks)*

  ii the size of the smallest angle in the triangle. *(4 marks)*

6 Calculate the coefficient of x^4 in the expansion, in powers of x, of $(2 + x)^6$. *(3 marks)*

In the expansion, in powers of x, of $(1 + ax)(2 + x)^6$, the coefficient of x^5 is 18.
Calculate the value of a. *(5 marks)*

7 **a** Evaluate $\int_1^4 \left(x + \dfrac{1}{x}\right)^2 dx$. *(4 marks)*

  b The gradient at the point (x, y) on a curve is given by $\dfrac{dy}{dx} = 4x - 3$.

    The curve passes through the point (2, 1). Find the equation of the curve. *(5 marks)*

8

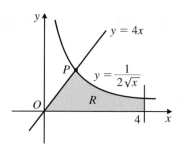

The line $y = 4x$ intersects the curve $y = \dfrac{1}{2\sqrt{x}}$ at the point P. The region R is

enclosed by the line $y = 4x$, the curve $y = \dfrac{1}{2\sqrt{x}}$, the x-axis, and the line $x = 4$

(see diagram).

i Show that the coordinates of the point P are $\left(\frac{1}{4}, 1\right)$. *(2 marks)*

ii Find the area of R. *(8 marks)*

9 a Given that $\displaystyle\sum_{i=0}^{5} (1 + id) = 3$, find the value of d. *(4 marks)*

b A grandparent invests £3600 in a stakeholder pension plan for a child on the day the child is born. It is given that the rate of interest on the investment is 5% per annum. Show that this investment has increased in value to £8663.83 by the day of the child's eighteenth birthday. *(2 marks)*

It is assumed that additional investments of £3600 are made on every birthday of the child, until the last investment is made on the eighteenth birthday. The money remains invested until the child reaches the age of fifty. Given that the rate of interest remains 5% per annum, calculate the total value of the investments on the child's fiftieth birthday, to the nearest £. *(8 marks)*

Answers

SKILLS CHECK 1A (page 5)

1 7.8 cm
2 15.8 cm (3 s.f.)
3 **a** 24.5° (3 s.f.)   **b** 11.0 cm² (3 s.f.)
4 **a** 52.1° (3 s.f.)   **b** 92.9° (3 s.f.)
5 **a** 8.25 cm (3 s.f.)   **b** 35.8 cm² (3 s.f.)
6 **b** **i** 11.6 cm² (3 s.f.), 17.7 cm² (3 s.f.)
       ii 4.10 cm (3 s.f.), 9.19 cm (3 s.f.)
7 55.9° (3 s.f.)
8 **a** 19.2 cm (3 s.f.)   **b** 69.2 cm² (3 s.f.)
9 **a** 17.9 m (3 s.f.)   **b** 57.1° (3 s.f.)
   c 60.9° (3 s.f.)   **d** 265 m² (3 s.f.)
10 **a** 30°, 6.46 cm (3 s.f.)   **b** 19.3 cm (3 s.f.)
   c $\sin A = 0.5 \Rightarrow a = 30°$, 150° giving two triangles with same base and same perpendicular height (see CD for diagrams).

SKILLS CHECK 1B (page 8)

1 **a** 4.89 radians (3 s.f.)   **b** 85.9° (3 s.f.)
2 **a** 120°   **b** 135°   **c** 270°   **d** 105°
3 **a** $\frac{1}{4}\pi$ **b** $\frac{5}{6}\pi$ **c** $\frac{11}{6}\pi$ **d** $\frac{4}{3}\pi$
4 **a** 3 cm   **b** 13 cm
   c 7.06 cm² (3 s.f.)   **d** 7.5 cm²
5 **a** 1.2 radians   **b** 64.9 cm² (3 s.f.)
6 **a** 1.5 radians   **b** 21 cm
7 **a** 0.848   **b** 27.1 cm² (3 s.f.)   **c** 3.14 cm² (3 s.f.)
8 **a** $6 + 17\theta$ **c** 12.75 cm²
9 **a** 4.84 cm (3 s.f.)   **b** 1.05 cm (3 s.f.)   **c** 16.9 cm (3 s.f.)
   d 13.5 cm² (3 s.f.)   **e** 1.05 cm² (3 s.f.)   **f** 12.5 cm² (3 s.f.)
10 **a** 4.57 cm (3 s.f.)   **b** 1.34 cm² (3 s.f.)

SKILLS CHECK 1C (page 16)

1 **a** 17°, 163° (nearest °)   **b** 60°, 300°
   c 124°, 304° (nearest °) **d** 10°, 110°, 130°, 230°, 250°, 350°
   e 105°, 165°, 285°, 345° **f** 90°
2 **a** $\frac{1}{3}\pi, \frac{2}{3}\pi$ **b** $\frac{3}{4}\pi$ **c** $\frac{2}{3}\pi$
   d $\frac{7}{12}\pi, \frac{11}{12}\pi$ **e** $\frac{1}{9}\pi, \frac{5}{9}\pi, \frac{7}{9}\pi$ **f** $\frac{1}{24}\pi, \frac{7}{24}\pi, \frac{13}{24}\pi, \frac{19}{24}\pi$
3 ±0.72 radians, ±5.56 radians (2 d.p.)
4 −90°, −30°, 30°, 90°
5 −360°, −315°, −225°, −180°, 0°, 45°, 135°, 180°, 360°
7 $\frac{4}{3}$
8 **a** 1   **b** $\frac{1}{12}\pi, \frac{5}{12}\pi, \frac{3}{4}\pi$ [0.26^c (2 d.p.), 1.31^c (2 d.p.), 2.36^c (2 d.p.)]
9 $\frac{1}{2}\pi, \frac{11}{6}\pi$ [1.57^c (2 d.p.), 5.76^c (2 d.p.)]
10 **b** 0°, 120°, 240°, 360°   **c** 0°, 60°, 120°, 180°

Exam practice 1 (page 16)

1 **i** 77 m (nearest m)   **ii** 874 m² (3 s.f.)
2 **i** 60 cm   **ii** 98.6 cm (nearest mm)
3 **i** 0.925 rad (3 d.p.)   **ii** 40 cm (2 s.f.)
4 **i** 2.3   **ii** 460 cm²
5 **i** $\frac{5}{3}\pi$ **ii** 109 cm (3 s.f.)   **iii** 589 cm² (3 s.f.)
6 $\dfrac{3\sqrt{3}}{2}$
7 240, 300
8 **a** $\frac{7}{10}\sqrt{2}$ **b** **ii** 0°, 30°, 150°, 180°, 360°
9 **i** $3 - 3\sin^2\theta - 2\sin\theta$ **ii** 19.5 (1 d.p.), 160.5 (1 d.p.), 270
10 **ii** 9.2° (1 d.p.), 99.2° (1 d.p.), 189.2° (1 d.p.), 279.2° (1 d.p.)
11 1.49^c (2 d.p.), 5.59^c (2 d.p.)
12 120°, 300°
13 **i** $-3\sin^2\theta - \sin\theta + 2, a = -3, b = -1, c = 2$
   ii 0°, 180°, 199.5° (1 d.p.), 340.5° (1 d.p.)

SKILLS CHECK 2A (page 21)

1 0, 3, 8, 15
2 1, 1.3333..., 1.5555..., 1.7037...
3 **a** 4, 8, 16, 32   **b** 7, 15, 31, 63
4 **a** $x_{n+1} = 0.9x_n$ **b** £5900
5 10, 3.1623, 1.7783, 1.3335, 1.1548, 1.0746, 1.0366
6 −4.833..., −5.034..., −4.993..., −5.001..., −4.999..., −5.000...
7 **a** 78   **b** 31.5
8 **a** 42   **b** 15
9 −4, −9, −7
10 **a** 40   **b** $2\frac{103}{210}$

SKILLS CHECK 2B (page 26)

1 **a** 5, 1190   **b** −3, −610
2 $u_n = n - \frac{1}{2}, S_{200} = 20\,000$
3 **a** 3   **b** $3n + 5$ **c** 1100
4 **a** 3   **b** $a = 9, d = -1$ **c** −165
5 **a** £50   **b** £10 500
6 **a** 741   **b** 1179
7 99
8 9
9 **a** 8, 10, 12   **b** 2   **c** 20   **d** 540
10 **a** 15 months   **b** £725

SKILLS CHECK 2C (page 29)

1 **a** **i** 31 250   **ii** 39 062
   b **i** $\frac{7}{64}$ **ii** $13\frac{57}{64}$ **iii** 14
   c **i** 64   **ii** 43
2 −59 048
3 £123 000
4 **a** $a = 40, r = \frac{1}{4}$ **b** $53\frac{1}{3}$
5 **a** $-\frac{1}{3}$ **b** $\frac{1}{324}$
6 **a** 699 048   **b** 3
7 137 262
8 **a** £512   **b** £1023
9 **a** 16   **b** $p = 14, q = 6$
10 1

SKILLS CHECK 2D (page 33)

1 $16 - 96x + 216x^2 - 216x^3 + 81x^4$
2 $1 + 28y + 336y^2 + 2240y^3$
3 **a** $243 - 810x + 1080x^2$ **b** $A = 1215, B = -3564, C = 3780$
4 **a** $1 + 12x + 60x^2 + 160x^3$ **b** 1.1262 (4 d.p.)
5 0.972 333 8 (7 d.p.)
6 $40\sqrt{6}$
7 4
8 **a** $n = 8, k = -\frac{1}{2}$ **b** $-7x^3$
9 **a** $1 + 6ax + 15a^2x^2$ **b** $a = 3, b = 2$
10 **a** $1 - 15x + 90x^2 - 270x^3$ **b** −180

Exam practice 2 (page 33)

1 80 200
2 **i** 0.6   **ii** 81 850
3 **i** 1.08   **ii** 12.5
4 **i** 0.432 (3 s.f.)   **ii** 148 (3 s.f.)   **iii** 150
5 3135
6 96
7 $a = 3, r = \frac{1}{4}; 2.4$
8 15 050
9 **i** 1905   **iii** 997 minutes
10 **i** $16 + 96x + 216x^2 + 216x^3 + 81x^4$ **ii** 124
11 **i** $1 + 7x + 21x^2 + 35x^3$ **ii** 1071

SKILLS CHECK 3A (page 37)

1 a $\frac{1}{2}, 2, -\frac{1}{3}$ b $\pm 2, \pm 3$ c $\frac{1}{4}, -\frac{4}{3}, 2$ d $-3, 1$
2 a $1, 2, -3$ b $3, -2 \pm \sqrt{2}$ c $-1, -2, -3$
3 a -1 b 8 c $0, (x-1)$ is a factor
4 a $x^2 - 3x - 5 - \dfrac{18}{x-3}$ b $x^2 + 5x - 6 + \dfrac{7}{x+1}$

   c $2x^2 + 5x + 6 + \dfrac{17}{x-2}$
5   $2, 2 \pm \sqrt{11}$
6   $(x+1)^3$
7   $a = 2, b = 4$
8   $a = 4, b = -1$

SKILLS CHECK 3B (page 41)

1 a 3 b 2 c 3 d $\frac{1}{2}$
2 a 9 b -2 c $\frac{3}{2}$ d $\frac{3}{2}$
3 a 2 b 36
4 a 5 b $-\frac{1}{2}$ c 3
5   $\log_2 p + \frac{1}{2}\log_2 q - \frac{1}{2} - \frac{3}{2}\log_2 r$
6   36
7 a i 1 ii 3 b -3
8 a 4.75 (3 s.f.)   b $x \geqslant 0.945$ (3 s.f.)   c $x > 4.32$ (3 s.f.)
9 a $a = d = \log_a b$ b 55
10 b $x = 0, 2$

Exam practice 3 (page 42)

1   3
2 i 2 ii $2, \sqrt{2}, -\sqrt{2}$
3   $(x-1)(x-2)(2x+3)$
4 i $a = 2, b = -5, c = 3; (x-1)(2x^2 - 5x + 3)$ ii $1, -0.5, 3$
5 i -8 ii -12
6 ii $p = -\sqrt{5}, q = \sqrt{5}$
7   $a = -1, b = 2$
8   $\log_2\left(\dfrac{x+2}{x}\right); \frac{2}{7}$
9 a 3 b 3
10   7.1 (1 d.p.)
11 ii $x = \frac{1}{2}$
12 i 2^{4x+2}
13   183

SKILLS CHECK 4A (page 46)

1 a $\frac{4}{3}x^3 + c$ b $\frac{2}{15}z^3 + c$ c $-6y + c$
2 a $y = \frac{1}{6}x^6 + x^3 + c$ b $y = \frac{2}{5}x^5 - \frac{1}{8}x^4 + c$ c $y = \frac{2}{3}x^3 + \frac{7}{2}x^2 + c$

   d $y = -\frac{3}{2}x^2 + 4x + c$ e $y = -\frac{1}{10}x^4 - x + c$ f $y = 2cx + k$
3 a $2x^2 - 5x - 12$ b $\frac{2}{3}x^3 - \frac{5}{2}x^2 - 12x + c$
4 a $A = \frac{1}{2}t^8 + c$ b 127.5
5   $y = \frac{5}{2}x^2 - 3x$
6 a $-\dfrac{3}{x} + c$ b $6\sqrt{x} + c$ c $\frac{4}{5}t^{\frac{5}{4}} + c$
7 a $x^{\frac{5}{2}}$ b $\frac{2}{7}x^{\frac{7}{2}} + c$
8 a $\frac{1}{3}x^3 + \frac{2}{3}x^{\frac{3}{2}} + c$ b $\frac{2}{3}x^{\frac{3}{2}} + 4x^{\frac{1}{2}} + c$

9   $y = \sqrt{x} + 2$
10   $x^{\frac{3}{2}} - x + 2$

SKILLS CHECK 4B (page 52)

1 a 7 b 6.4 c 1.25 d 1
2 a 12 b -5.25
3 a $\frac{1}{2}$ b $-\frac{3}{8}$
4 a $x + x^{-2}$ b 2
5   16
6 a $x = 1, y = 3; x = -2, y = 6$
   b i $P(-2, 6), Q(1, 3)$ ii $4\frac{1}{2}$
7   $\frac{3}{8}$
8 b 0.25

SKILLS CHECK 4C (page 54)

1 b 22.5 c Underestimate
2   0.406 (3 s.f.)
3 a $p = 0.825, q = 0.540$ b 0.84 (2 d.p.)
4 a Strips have width 0.4 b 0.559 (3 s.f.)
5 b 39.8 (3 s.f.)
6   24.3 (3 s.f.)
7 a 10 b $10\frac{2}{3}, 6.25\%$
8 b $30\frac{2}{3}$ c Split into more intervals
9 a 51.25 b 0.655 (3 s.f.)
10 a $1, 0.93, 0.76, 0.58, 0.44, 0.33$ b 1.35 (3 s.f.)

Exam practice 4 (page 55)

1   $\frac{1}{4}x^4 + x^2 - \dfrac{1}{x} + c$
2   $y = -x^2 + 3x + 5$
3 i $\frac{1}{3}x^3 + \frac{1}{2}x^2 + c$ ii $-\dfrac{1}{x} + c$
4 a 32 b $\frac{1}{8}$
5   $y = x^3 - \frac{1}{2}x^2 - 5$
6 i $y = 2x^{\frac{7}{2}} - x^3 + c$ ii $y = 2x^{\frac{7}{2}} - x^3 + 4$
7 i $k = 3$ ii $2x^{\frac{3}{2}} - x^3 + 41x + 18$
8 a $\frac{1}{3}x^3 + 2x + c$ b i $-\dfrac{3}{x} + c$ ii 3
9   0.5
10 ii They are equal
11 i $(-1, 6), (\frac{3}{2}, \frac{7}{2})$ ii $x < -1, x > \frac{3}{2}$
12   $-1, 4; 20\frac{5}{6}$
13 i 1.09 (3 s.f.)   ii overestimate
14   2.2

Practice exam paper (page 58)

1 ii $(x+2)(2x^2 - 3x + 5)$
2   4
3   $-1 + \sqrt{2}, 65.5°$ (1 d.p.), $294.5°$ (1 d.p.)
4 i $\dfrac{\pi}{6}$ ii 0.425 cm^2
5 i 2.65 cm (3 s.f.)   ii $40.9°$ (3 s.f.)
6   $60, 0.1$
7 a 27.75 b $y = 2x^2 - 3x - 1$
8 ii 1.625
9 a -0.2 b £523 860

SINGLE USER LICENCE AGREEMENT FOR CORE 2 FOR OCR CD-ROM
IMPORTANT: READ CAREFULLY

WARNING: BY OPENING THE PACKAGE YOU AGREE TO BE BOUND BY THE TERMS OF THE LICENCE AGREEMENT BELOW.

This is a legally binding agreement between You (the user or purchaser) and Pearson Education Limited. By retaining this licence, any software media or accompanying written materials or carrying out any of the permitted activities You agree to be bound by the terms of the licence agreement below.

If You do not agree to these terms then promptly return the entire publication (this licence and all software, written materials, packaging and any other components received with it) with Your sales receipt to Your supplier for a full refund.

YOU ARE PERMITTED TO:

- Use (load into temporary memory or permanent storage) a single copy of the software on only one computer at a time. If this computer is linked to a network then the software may only be used in a manner such that it is not accessible to other machines on the network.

- Transfer the software from one computer to another provided that you only use it on one computer at a time.

- Print a single copy of any PDF file from the CD-ROM for the sole use of the user.

YOU MAY NOT:

- Rent or lease the software or any part of the publication.

- Copy any part of the documentation, except where specifically indicated otherwise.

- Make copies of the software, other than for backup purposes.

- Reverse engineer, decompile or disassemble the software.

- Use the software on more than one computer at a time.

- Install the software on any networked computer in a way that could allow access to it from more than one machine on the network.

- Use the software in any way not specified above without the prior written consent of Pearson Education Limited.

- Print off multiple copies of any PDF file.

ONE COPY ONLY

This licence is for a single user copy of the software

PEARSON EDUCATION LIMITED RESERVES THE RIGHT TO TERMINATE THIS LICENCE BY WRITTEN NOTICE AND TO TAKE ACTION TO RECOVER ANY DAMAGES SUFFERED BY PEARSON EDUCATION LIMITED IF YOU BREACH ANY PROVISION OF THIS AGREEMENT.

Pearson Education Limited and/or its licensors own the software.
You only own the disk on which the software is supplied.

Pearson Education Limited warrants that the diskette or CD-ROM on which the software is supplied is free from defects in materials and workmanship under normal use for ninety (90) days from the date You receive it. This warranty is limited to You and is not transferable. Pearson Education Limited does not warrant that the functions of the software meet Your requirements or that the media is compatible with any computer system on which it is used or that the operation of the software will be unlimited or error free.

You assume responsibility for selecting the software to achieve Your intended results and for the installation of, the use of and the results obtained from the software. The entire liability of Pearson Education Limited and its suppliers and your only remedy shall be replacement free of charge of the components that do not meet this warranty.

This limited warranty is void if any damage has resulted from accident, abuse, misapplication, service or modification by someone other than Pearson Education Limited. In no event shall Pearson Education Limited or its suppliers be liable for any damages whatsoever arising out of installation of the software, even if advised of the possibility of such damages. Pearson Education Limited will not be liable for any loss or damage of any nature suffered by any party as a result of reliance upon or reproduction of or any errors in the content of the publication.

Pearson Education Limited does not limit its liability for death or personal injury caused by its negligence.

This licence agreement shall be governed by and interpreted and construed in accordance with English law.